THE PRINCIPLES OF
CLOUD-CHAMBER TECHNIQUE

BY

J. G. WILSON
Senior Lecturer in Physics in the University of Manchester

CAMBRIDGE
AT THE UNIVERSITY PRESS
1951

CAMBRIDGE
UNIVERSITY PRESS

University Printing House, Cambridge CB2 8BS, United Kingdom

Published in the United States of America by Cambridge University Press, New York

Cambridge University Press is part of the University of Cambridge.

It furthers the University's mission by disseminating knowledge in the pursuit of education, learning and research at the highest international levels of excellence.

www.cambridge.org
Information on this title: www.cambridge.org/9781107680890

© Cambridge University Press 1951

First published 1951
First paperback edition 2014

A catalogue record for this publication is available from the British Library

ISBN 978-1-107-68089-0 Paperback

CAMBRIDGE MONOGRAPHS ON PHYSICS

GENERAL EDITORS

N. FEATHER, F.R.S.
Professor of Natural Philosophy in the University of Edinburgh

D. SHOENBERG, PH.D.
Fellow of Gonville and Caius College, Cambridge

THE PRINCIPLES OF
CLOUD-CHAMBER TECHNIQUE

GENERAL PREFACE

The Cambridge Physical Tracts, out of which this series of Monographs has developed, were planned and originally published in a period when book production was a fairly rapid process. Unfortunately, that is no longer so, and to meet the new situation a change of title and a slight change of emphasis have been decided on. The major aim of the series will still be the presentation of the results of recent research, but individual volumes will be somewhat more substantial, and more comprehensive in scope, than were the volumes of the older series. This will be true, in many cases, of new editions of the Tracts, as these are re-published in the expanded series, and it will be true in most cases of the Monographs which have been written since the War or are still to be written.

The aim will be that the series as a whole shall remain representative of the entire field of pure physics, but it will occasion no surprise if, during the next few years, the subject of nuclear physics claims a large share of attention. Only in this way can justice be done to the enormous advances in this field of research over the War years.

N. F.
D. S.

CONTENTS

AUTHOR'S PREFACE

While preparing this book, I have had chiefly in mind the requirements of workers who use or plan to use cloud chambers, and those who may be helped by an estimate of the potentialities of the cloud chamber as an instrument of precise measurement. I have as far as possible avoided discussion of details of construction or manipulation, for there is little evidence of any correlation between the diversity of practice which now exists and the quality of performance that results from it, and I have included chapters dealing with the condensation process, ionization, and the general problem of the establishment of supersaturation and its persistence, which are the basic mechanisms upon which any cloud-track apparatus must be developed. I have also dealt in some detail with the technique of measurement, and here the majority of examples given refer to cosmic-ray investigations, for it was in this field that most of the refinements of measuring technique first came to be of major importance.

It is not easy to attribute a great deal of undocumented information to its correct source, but I must acknowledge with gratitude many discussions on all aspects of cloud-chamber technique over the last fifteen years with Professors C. T. R. Wilson, P. M. S. Blackett and the late E. J. Williams. I am also indebted to Professors P. I. Dee and S. Gorodetsky and to Dr G. R. Evans for the benefit of their experience on specific topics, and to my colleagues in Manchester, in particular to Drs G. D. Rochester and C. C. Butler, with whom I have had the opportunity of discussing the interpretation of photographs, a topic of rapidly increasing importance.

J. G. WILSON

Manchester
AUGUST 1948

CHAPTER I

THE CONDENSATION AND GROWTH
OF DROPS

1.1. Condensation on nuclei

Attention was first drawn to the necessity of nuclei being present, upon which drops of liquid tend to form, before condensation will take place in a supersaturated vapour, by the researches of Coulier (1875) and Aitken (1880–1), who showed that the cloud formed in damp air following slight supersaturation was made more dense in the presence of combustion products, and was reduced by filtering the air through cotton-wool or by repeated cloud formation in an enclosed gas volume. The nuclei of this condensation are dust particles, some visible in an intense beam of light, many invisible, but all capable of being removed from an enclosed mass of gas. The resulting clean gas will sustain a considerable adiabatic expansion without drop formation in the body of the gas.

Coulier enclosed air in a flask together with water and produced supersaturation by compressing a hollow india-rubber ball connected to the flask and then suddenly releasing it.* He found that dense condensation followed a small expansion in fresh, unfiltered air but that after the enclosed air had stood for some days, the condensation following such an expansion was negligible. Similarly, if the entering air were filtered through cotton-wool it was from the beginning in the state from which little condensation took place when slight supersaturation was established.

In his earlier experiments, Aitken produced supersaturation by blowing steam into a large vessel containing the gas to be tested; later, he also used an expansion method not unlike that described by Coulier, whose work was, at the time, unknown to him. He showed that filtered air was not effective in promoting condensation and that enclosed air might be cleaned by a succession of cloud-forming expansions in which the drops formed were allowed to fall to the bottom of the containing vessel. Aitken constructed apparatus

---

* An instrument made by C. T. Knipp and similar to that described by Coulier is exhibited at the Cavendish Laboratory. An α-particle source is enclosed, and by appropriate pressure on the ball, condensation on α tracks can be achieved.

for measuring the number of dust particles capable of acting as nuclei of condensation in samples of air, and undertook an extensive survey of the dust content of natural air. This usually varied from $\sim 500/\text{cm}^3$ remote from human habitation (the lowest figure recorded is about $50/\text{cm}^3$) to $\sim 10^6/\text{cm}^3$ in city conditions and to still greater values indoors in the presence of flames. He established the existence of specific natural sources of nuclei, in particular (1911) the foreshore when illuminated in direct sunshine.

Aitken further showed that 'dust' nuclei are not equally effective, that condensation takes place on some for an expansion ratio (water vapour in air) as low as 1·004, while some require an expansion ratio as great as 1·02. He was not able completely to satisfy himself, however, as to whether, when all of these have been removed, condensation could still be initiated at very high expansions.

The conditions for condensation in a clean damp gas were demonstrated in the classical experiments of C. T. R. Wilson (1897, 1899), who distinguished two critical values of supersaturation corresponding to onset limits of effectiveness of differing nuclei. The lower of these, at roughly fourfold supersaturation for water vapour in air, was shown to involve gaseous ions as nuclei; Wilson described this as the limit of rain-like condensation, for under ordinary conditions the number of cloud droplets formed was small enough for these to be individually prominent. The upper limit, at about eightfold supersaturation, was described as the limit of cloud-like condensation; this was not a sharply defined condition, and the number of nuclei effective increased rapidly and continuously with increasing supersaturation. The nuclei responsible for this increase were probably uncharged aggregates of a very few molecules. Following the formation of cloud and its subsequent evaporation, nuclei effective at very low supersaturations were found to be present. These were permanently removed by repeated cloud formation in the same way as were the Aitken dust particles.

Hence, in dust-free gas, three main groups of condensation nuclei were distinguished:

    (a) nuclei derived from gaseous ions (charged nuclei),

    (b) uncharged small nuclei, probably aggregates of a few molecules,

    (c) re-evaporation nuclei, effective at very small supersaturations.

The purpose of the cloud chamber is to study the motion of ionizing particles from records of the drops condensed on ions formed along the trajectories followed by these particles, and to do this, each type of condensation must to some degree be controlled. Drops are wanted solely on recently produced charged nuclei.

The nuclei from gaseous ions are significant only so long as they may be sufficiently identified with a particular particle trajectory. Old nuclei, which have drifted far from the points at which they were formed, are therefore removed from the experimental gas volume by an electrostatic field (the 'clearing' or 'sweeping' field).

Uncharged small nuclei are always present in a gas, and cloud-chamber operation is only possible in a combination of gas and vapour for which the onset of condensation on these nuclei takes place at an appreciably higher supersaturation than that necessary for condensation on ions. This condition is met by many combinations, but requires the rigorous exclusion of certain impurities.

Condensation on re-evaporation nuclei, if present, cannot be avoided. These, which may be permanently removed, are therefore swept away from the chamber by a series of subsidiary 'cleaning' expansions following each full expansion to the ion limit.

In the following paragraphs we discuss the features of condensation on nuclei, along lines due to J. J. Thomson (1888). This is essentially a descriptive treatment in terms of the properties of macroscopic drops, but it is here applied to very small drops, of radius approaching the fundamental limit set by the average distance between molecules in the liquid phase. For a more complete treatment of these very small drops, we refer to Frenkel (1946, Chap. VII); but it may be noted here that the unsatisfactory introduction in the following sections of postulated variations of surface tension with drop size is unlikely to be significant in regions for which the concept of surface tension is valuable.

1.2. The stability of charged drops

The surface energy of the drop, radius a, surface tension T, in an external medium of dielectric constant, ϵ_1, is

$$4\pi a^2 T + \frac{1}{2}\frac{q^2}{\epsilon_1 a},$$

and a change of radius, δa, leads to a change of energy

$$\frac{d}{da}\left(4\pi a^2 T + \frac{1}{2}\frac{q^2}{\epsilon_1 a}\right)\delta a,$$

which would not take place in the corresponding evaporation or condensation at a plane surface. This change of energy (with decrease of radius) may be equated to the work needed to bring the amount of vapour produced by evaporation at the vapour pressure, p, in equilibrium with the drop, to the saturation pressure, p_0, over a plane surface. Thus

$$\frac{d}{da}\left(4\pi a^2 T + \frac{1}{2}\frac{q^2}{\epsilon_1 a}\right)\delta a = 4\pi a^2 \sigma \delta a \frac{R\theta}{M}\ln\frac{p}{p_0},$$

where R is the gas constant, M the molecular weight of the vapour, θ the absolute temperature, and σ the density of the drop, or

$$\frac{R\theta\sigma}{M}\ln\frac{p}{p_0} = \frac{2T}{a} + \frac{dT}{da} - \frac{q^2}{8\pi\epsilon_1 a^4}. \tag{1}$$

We are concerned, however, in practice, with drops carrying a single electronic charge, and this can only lead to an effective surface charge distribution by polarizing the drop. The effective charge will now depend on ϵ_2, the dielectric constant of the condensed liquid, and Tohmfer and Volmer (1938) have modified equation (1) to the form

$$\frac{R\theta\sigma}{M}\ln\frac{p}{p_0} = \frac{2T}{a} + \frac{dT}{da} - \left(\frac{1}{\epsilon_1} - \frac{1}{\epsilon_2}\right)\frac{q^2}{8\pi a^4}, \tag{2}$$

where, however, as has been pointed out by Tohmfer and Volmer and by Glosios (1939), ϵ_1 and ϵ_2 may differ appreciably from the bulk values.

If, in (1) or (2), we can assume that the surface tension, T, is independent of drop radius, the variation of the equilibrium supersaturation, (p/p_0), with drop radius is represented by the full curve shown schematically in fig. 1, where

$$a_0^3 = \frac{q^2}{16\pi T}\left(\frac{1}{\epsilon_1} - \frac{1}{\epsilon_2}\right),$$

and

$$a_1^3 = \frac{q^2}{4\pi T}\left(\frac{1}{\epsilon_1} - \frac{1}{\epsilon_2}\right).$$

The curve divides the space of the diagram, in which each point represents the condition of a drop of given radius in vapour of given pressure, into two halves: points above the curve represent drops surrounded by vapour which will condense upon them, and this is therefore a domain of drop growth; similarly, the region below the curve is a domain of evaporation. We postpone a discussion of the evaporation of drops and observe that two categories of growth occur, according as the growth of the drop does or does not bring the representative point on to the curve of fig. 1. If condensation on the drop leads to a point on the curve, then growth stops while the drop is still invisible (A of fig. 1); if it does not (B), then

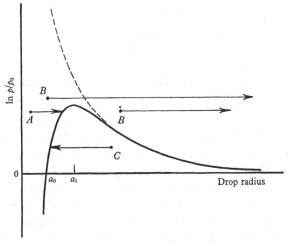

Fig. 1. Vapour-pressure equilibrium for small droplets (full curve, charged; broken line, uncharged).

the drop grows indefinitely until external factors remove the condition of supersaturation. In the cloud chamber no drops of category B exist initially, but at an adiabatic expansion the drops of category A which are present pass over into category B if the supersaturation obtained is greater than the equilibrium value for radius a_1.

The orders of magnitude concerned are readily derived; for water at 0° C. we find for a drop carrying a single electronic charge

$$\left.\begin{array}{l} a_1 = 6 \times 10^{-8}\,\text{cm.} \\ (p/p_0)_{a_1} = 4 \cdot 2 \end{array}\right\} \tag{3}$$

the latter quantity corresponding closely to the experimental supersaturation limit for C. T. R. Wilson's 'rain-like' condensation.

1.3. Growth from charged nuclei

To the first approximation, ions formed in damp gas will become points of aggregation for a water droplet of radius about a_0 in saturated vapour and rather less in unsaturated vapour, a droplet of rather less than ten molecules. If supersaturation is now established in the surrounding vapour, these charged droplets grow, following the increasing equilibrium radius until the supersaturation $(p/p_0)_{a_1}$ is reached. Then all of these drops grow to visible size.

In detail it is found

(1) that all charged nuclei of a given sign do not become effective at exactly the same supersaturation,

(2) that nuclei of opposite sign of charge require different supersaturations to initiate indefinite growth.

These features fit in well with the model we have considered. The critical radius, a_1, corresponds, for water, to a droplet of about 30 molecules. The energy increment per molecule is small enough for appreciable fluctuations of this number to occur, and hence at any instant the size of charged droplets in equilibrium will form a statistical distribution about an average size of about 30 molecules. As regards the sign of charge, equation (2) is clearly incomplete when, as for water, strongly polar molecules form an orientated surface layer. No detailed treatment of this modification has yet been made. Some typical values given by Scharrer (1939) of the critical supersaturation for drop growth from charged droplets are shown in Table I; $(p/p_0)_{\text{theor.}}$ refers to equation (1) and $(p/p_0)_{\text{exp.}}$ is the value for the sign of charge on which condensation takes place more easily; when both signs are indicated the limits were not separated.

TABLE I

Vapour	M	σ g./cm.^{3}	T dyne./cm.	$(p/p_0)_{\text{theor.}}$	$(p/p_0)_{\text{exp.}}$	Sign of ion
H_2O	18	1·00	76·5	4·46	4·14	−
C_2H_5OH	46	0·81	23·1	2·52	1·94	+
CH_3OH	32	0·81	24·7	2·07	2·95	+
C_6H_6	78	0·92	33·3	11·40	4·94	+, −
CCl_4	154	1·63	31·1	12·70	6·0	+, −
$CHCl_3$	119	1·50	30·3	7·4	3·45	+
C_6H_5Cl	112	1·15	38·0	30·0	8·9	−

1.4. Condensation on uncharged nuclei

For uncharged nuclei, equations (1) and (2) become

$$\frac{R\theta\sigma}{M}\ln\frac{p}{p_0} = \frac{2T}{a} + \frac{dT}{da}, \tag{4}$$

and, in the absence of particular assumptions about the variation of T with drop size, no mechanism defining a critical supersaturation comparable with that for charged drops exists. Experimentally, however, condensation does take place on uncharged nuclei, rapidly increasing numbers of which become effective as the supersaturation is increased. C. T. R. Wilson described the cloud limit for condensation of water in air, when so many nuclei were effective that a continuous cloud was formed in the chamber, at about eightfold supersaturation. He also showed that this limit of supersaturation is the same in such different gases as oxygen, hydrogen and carbon dioxide, and it is thus probable that the nuclei concerned do not involve molecules of the permanent gas.

We consider these nuclei, therefore, to be uncharged aggregates of vapour molecules forming part of a statistical distribution like that for charged aggregates but with the difference that zero aggregation is the most probable condition, and that while condensation is in practice required on at least a large fraction of charged nuclei, only very rare and extreme fluctuations of uncharged aggregates ever reach the size leading to continued growth.

The statistical equilibrium of such nuclei of the liquid phase in supersaturated vapour has been treated by Volmer and Weber (1926), Farkas (1927) and others, and yields an estimate of the supersaturation at which appreciable numbers of nuclei will become big enough to act as condensation nuclei and grow to visible size. Taking a constant value of surface tension, the number of nuclei, $Z(a)$, reaching radius a, per cm.[3] per sec., is given by Volmer and Flood (1934) in the form

$$Z(a) = C(\alpha p_0)\frac{M}{N\sigma}\sqrt{\left(\frac{T}{k\theta}\right)}\exp\left[-\frac{4}{3}\frac{\pi a^2 T}{k\theta}\right], \tag{5}$$

where C is an undetermined constant, (αp_0) is the mean number of molecule impacts/cm.[2] sec., N the Avagadro number, and k the Boltzmann constant.

A value of $Z(a)$ is now adopted to define the effective onset of condensation on uncharged nuclei. This has been taken by Volmer and Weber and by subsequent workers at about $Z(a) = 1$, which is considered to be of the correct order for the first signs of condensation in a normal cloud chamber but which clearly represents a very much lower density of drop formation than C. T. R. Wilson's original criterion. Scharrer, for example, has measured the onset limit when the density of drops in his chamber was 1–5 drops/cm.[3] The critical supersaturation then follows if the value of a defined by equation (5), where $Z(a) = 1$, is used in equation (4).

Thus
$$R\theta \frac{\sigma}{M} \ln\left(\frac{p}{p_0}\right)_{\text{crit.}} = \frac{2T}{a},$$

$$\frac{4}{3}\frac{\pi a^2 T}{k\theta} = \ln\left\{C(\alpha p_0)\frac{M}{N\sigma}\sqrt{\left(\frac{T}{k\theta}\right)}\right\},$$

and so
$$\ln\left(\frac{p}{p_0}\right)_{\text{crit.}} = \frac{4TM}{R\theta\sigma}\left(\frac{\pi T}{3k\theta}\right)^{\frac{1}{2}}\left[\ln\left\{C(\alpha p_0)\frac{M}{N\sigma}\sqrt{\left(\frac{T}{k\theta}\right)}\right\}\right]^{-\frac{1}{2}} \quad (6)$$

$$= K\left(\frac{T}{\theta}\right)^{\frac{3}{2}}\frac{M}{\sigma}, \quad (7)$$

if the small variations of the logarithmic term are ignored.

The undetermined constant in equation (5) has been estimated in different forms by Becker and Doring (1935) and Frenkel (1946). The problem is most complicated, for while, when supersaturation is established, the statistical equilibrium of nuclei of the liquid phase present in the initial saturated gas goes over, for the small aggregates, to a metastable equilibrium following closely a true equilibrium distribution, for larger aggregates virtually no progress is made towards the establishment of equilibrium, and the drops which grow to visible size are statistically quite negligible. The behaviour of the intermediate aggregates in the neighbourhood of the critical radius cannot be uniquely established. However, the absolute magnitude of the rate $Z(a)$ is not of particular value, for it clearly varies, among other factors, extremely rapidly with T. The problem of condensation on statistical aggregates which act as nuclei of the liquid phase has been treated in detail by Frenkel (1946, Chap. VII).

The above relation (equation (7)) of critical supersaturation with the bulk value of surface tension and with molecular volume was shown by Volmer and Flood (1934) to hold reasonably well as between water and several organic liquids, methyl alcohol alone exhibiting serious discrepancy. Their results, in which the theoretical and experimental values are fitted for water, are given in Table II. (The experimental values in the table are lower than those given by Wilson and the other early workers, for, as we have noted, they refer to a much lower density of condensation.)

TABLE II

Liquid	Temperature, θ, after expansion °K.	M	σ g./cm.^{3}	T dyne/cm.	$(p/p_0)_{\text{crit.}}$ Theor.	$(p/p_0)_{\text{crit.}}$ Exp.
Water	264	18·0	1·00	77·0	(4·85)	4·85
Methyl alcohol	270	32·0	0·81	24·8	1·84	3·20
Ethyl alcohol	273	46·0	0·81	24·0	2·30	2·34
n-Propyl alcohol	270	60·1	0·82	25·4	3·20	3·05
Isopropyl alcohol	265	60·1	0·81	23·1	2·90	2·80
Butyl alcohol	270	74·1	0·83	26·1	4·50	4·60
Nitromethane	252	61·0	1·20	40·6	6·25	6·05
Ethylacetate	240	88·1	0·94	30·8	11·0	12·3

1.5. Condensation in mixed vapours

That the relations between critical supersaturation, surface tension and molecular volume could be extended to condensation from mixed vapours was demonstrated by Flood (1934), who studied mixtures of ethyl alcohol and water. His results for this mixture, which is of great practical importance, are given in Table III, in which, as before, the theoretical expression is fitted to the experimental value for pure water. Flood's experimental results are also shown in fig. 2.

TABLE III

% of alcohol molecules in liquid	Temperature after expansion °K.	Effective M/σ cm.^{3}/g.mol.	T dyne/cm.	$(p/p_0)_{\text{crit.}}$ Theor.	$(p/p_0)_{\text{crit.}}$ Exp.
0	264	18·0	77·0	(4·85)	4·85
3·9	273	19·4	51·3	3·30	2·64
11·5	276	22·1	36·8	2·45	1·97
35·3	280	—	—	1·92	1·75
67·1	277	—	—	1·70	1·62
77·9	275	47·8	25·4	1·97	1·77
100	273	57·1	24·0	2·30	2·34

More recently, Flood's work has been extended by Beck (1941), who has investigated the variations of critical supersaturation for mixtures of the lower alcohols with water. In all cases a variation of the form described by Flood is found; while, in contrast, mixtures of alcohols show no condition of minimum critical supersaturation.

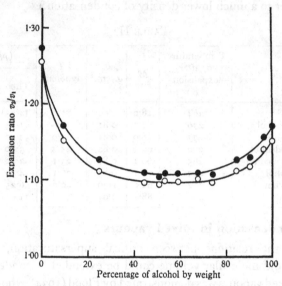

Fig. 2. Critical volume-expansion ratios for condensation on ions (○) and on background nuclei (●) in water-alcohol mixtures (Flood).

1.6. The practical onset of background condensation: contamination

It is found, for all suitable liquids, that the margin between the supersaturation at which substantially complete condensation on charged nuclei is obtained and that at which the drop density of condensation on uncharged nuclei becomes unworkable is always small. Moreover, in practice, the critical supersaturations for 'background' condensation on uncharged nuclei quoted above are found to be maximum values, the presence of contaminating material tending to depress the critical value towards and even below that for condensation on ions.

It is to be expected that any substance in small concentration, and hence present only in a small fraction of aggregates of molecules, affects the onset of background condensation, if at all, in the direction of depressing the limit of supersaturation, and practical experience bears out this conclusion. Our knowledge of substances which do lower the onset of background, and the treatment of apparatus so contaminated, is at present almost entirely empirical.

C. T. R. Wilson (1899) studied the effect of metal surfaces on the onset of condensation, using an experimental chamber divided into two parts, one of which contained the sample of material and the other, otherwise identical, acted as a control vessel. He found that the materials tested showed no tendency to produce nuclei effective in the range of low supersaturation which covered the condensation on 'dust' investigated by Aitken, but that they did produce in greater or less degree nuclei effective in the range $1\cdot252$ (ion onset) $< v_2/v_1$ (air, water) $< 1\cdot38$ (dense background cloud), or, in extreme cases, effective somewhat below the ion limit. Amalgamated zinc gave the largest amount of condensation, polished zinc or lead gave fewer drops and polished copper or tin very few. An old lead surface gave many fewer drops than a freshly scraped one. In hydrogen, only a very slight effect was observed for all metal surfaces.

The nuclei arising from metal surfaces are not affected by an electrostatic field, and from this feature Wilson was led to compare them with similar, field-insensitive nuclei, formed in oxygen and to a lesser extent in carbon dioxide, but not in hydrogen, by ultraviolet light. The latter nuclei were shown to be capable of continued growth to visible size under prolonged exposure without any expansion; they were found in oxygen and water vapour, both stringently purified and in vessels from which rubber and organic sealing materials were excluded, and in a convergent beam of ultraviolet light the seat of most intense formation and subsequent growth was shown to be in the region of focus and not near the illuminated containing walls. These properties lead Wilson to the conclusion that these nuclei were water droplets containing hydrogen peroxide as a non-volatile solute, the amount of which, in each droplet, increased during prolonged exposure to ultra-violet light with a resulting increase in the equilibrium diameter of the droplet (§ 1.2). He further considered it likely that the nuclei arising from the presence of metals might be of the same type, quoting in support the observation of Schönbein that hydrogen peroxide was produced when amalgamated zinc, water and oxygen were shaken together. When they arise from metal surfaces, such nuclei must be supposed to be denied the possibility of further growth, and the nuclei must be expected to develop on each individual hydrogen peroxide molecule. No significant extension of these observations seems since to have been made.

While 'trace' contaminations always depress the onset of background, it is possible that substances present in considerable concentration might enter a large fraction of the transitory condensation aggregates and shift the distribution so as to raise the onset limit. Beck (1941) has studied the addition of acetone to water-alcohol mixtures in concentrations up to 25 % of the whole liquid, and has concluded that it acts in this way. The effect, however, is certainly not large, and it is not yet clear that it is generally accepted. Alternatively, it is possible that if the contaminating material leading to premature background condensation were identified, some substance with properties analogous to a 'getter' in vacuum techniques could be introduced to sweep away the contamination. However, if the picture given by C. T. R. Wilson is correct, the hydrogen peroxide would only be vulnerable for a very short time between formation and the development of a drop nucleus upon it, and the 'getter' would have to be used as a coating on the suspect metal surfaces.

Some unpublished experiments by Dee* suggest that in normal chambers the first onset of background is, even in very clean conditions, controlled by 'trace' contamination, that is, perhaps, by peroxide nuclei. In this work a cloud chamber was constructed in which the initial adiabatic expansion was followed after a controlled interval by a small adiabatic compression. Thus supersaturation in the region of the critical condensation limits could be maintained for a short, known time and then removed. Dee found that for the condition of very thin background cloud, up to, say, 10 drops/cm.³, for which alone this mechanism altered the duration of conditions suitable for condensation on background to a significant extent, he was unable to detect any change of background

* I am indebted to Professor Dee for permission to quote this result.

density when the duration of supersaturation was varied. This result suggests that the 'life' of the nuclei responsible for the first onset of background is at least comparable with the sensitive time of the chamber concerned, say, about o·1 sec.

The detailed relation between the efficiency of condensation on ions and the onset of background condensation is of a practical importance that is not always appreciated. In a very clean chamber the appearance of condensation on the tracks of fast particles may be studied from the point at which condensation first begins up to that at which the density of background begins to increase seriously, and the improvement of brightness, strength and apparent contrast during the process is most striking. A rather larger amount of vapour becomes available for condensation, but much of the improvement arises from the increase in the number of ions condensed upon until the point is reached when all are effective. In a contaminated chamber, the premature onset of background merely prevents the use of the full expansion ratio possible in a clean chamber, and the less satisfactory tracks at a lower ratio are accordingly tolerated (see §4.2 below). It is extremely easy to allow a deterioration of quality to take place during an extended set of observations.

1.7. Re-evaporation nuclei

After an expansion leading to fresh condensation either on ions or uncharged nuclei, it is found that a cloud chamber contains nuclei on which further condensation will take place at very small supersaturation. These nuclei appear only to be produced by the evaporation of large droplets (hence we refer to them as 're-evaporation nuclei'), and are removed from the cloud chamber in the same way as dust nuclei. It is apparent that the course of evaporation indicated in the lower domain of fig. 1 is not followed in the re-evaporation from large droplets.

It was first suggested by Thomson that evaporating drops reached equilibrium at a comparatively large size ($a_2 > 10^{-7}$ cm.) determined by the variation of surface tension with drop radius (fig. 3). Here a_2 is given by

$$\frac{2T}{a} + \frac{dT}{da} = 0, \quad \frac{d^2T}{da^2} > \frac{2T}{a^2}. \tag{8}$$

If appreciable changes of surface tension do in fact occur for these larger sizes of droplet, the use of the bulk value of surface tension for the treatment of the small uncharged nuclei of background condensation must clearly be regarded as a crude approximation. Confirmation of the large size of these nuclei is given by the small but perceptible effect of a large electrostatic cleaning field in assisting the removal of some fraction of them.

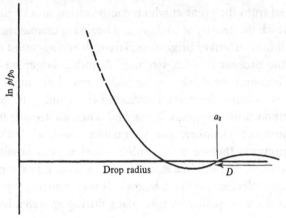

Fig. 3. Vapour-pressure equilibrium for evaporating drops; form of the equilibrium curve postulated to account for the formation of 're-evaporation nuclei'.

1.8. Rate of growth of drops

The way in which a drop grows in time from the condition B (fig. 1) is of considerable importance, as it bears upon the time of growth necessary before the condensed drop can be photographed. The problem is extremely complicated in detail: vapour is diffusing to the growing drop, while at the same time the heat of condensation which is being liberated at the drop is conducted away from it; the coefficients of diffusion and thermal conductivity vary from point to point through the relevant volume; moreover, as the drop grows, it begins to fall from its point of origin, and we may expect its rate of growth to be modified when this motion becomes large enough to bring the drop constantly into fresh undepleted vapour. We shall give only an elementary survey of the treatment of drop growth.

In the first approximation we assume the coefficient of diffusion to be a constant throughout the neighbourhood of a growing drop, and that the liberation of heat at the drop maintains the drop at some steady temperature. Then,* if D is the coefficient of diffusion of vapour in the gas of the chamber, ρ_2, ρ_0, ρ, respectively, the vapour density far from the drop, in equilibrium with it (ignoring the effect of curvature of surface for very small drops), and at distance a from its centre, the diffusion equation

$$\frac{\partial(\rho_2-\rho)}{\partial t} = \frac{D}{a}\frac{\partial^2 a(\rho_2-\rho)}{\partial a^2} \tag{9}$$

has the solution

$$(\rho_2-\rho) = (\rho_2-\rho_0)\left\{1-\frac{a_0}{a}+\frac{2}{\sqrt{\pi}}\frac{a_0}{a}\int_0^{\frac{a-a_0}{2\sqrt{(Dt)}}} e^{-x^2}dx\right\}, \tag{10}$$

where $a_0(t)$ is the instantaneous drop radius at time t. Then

$$\left(\frac{\partial\rho}{\partial a}\right)_{a=a_0} = (\rho_2-\rho_0)\frac{1}{a_0}\left(1+\frac{a_0}{\sqrt{(\pi Dt)}}\right), \tag{11}$$

and the growth of the drop is given by

$$\frac{a_0^2}{t} = \frac{2D(\rho_2-\rho_0)}{\sigma}\left(1+\frac{2}{\sqrt{(\pi D)}}\frac{a_0}{\sqrt{t}}\right), \tag{12}$$

where σ is the density of condensed liquid.

For water vapour condensing in air in typical cloud-chamber conditions,

$$D = 0\cdot20\,\text{cm.}^2\text{sec.}^{-1}, \quad (\rho_2-\rho_0) = 4\cdot10^{-6}\,\text{g.cm.}^{-3},$$
$$\sigma = 1\,\text{g.cm.}^{-3},$$

it is immediately clear that the second term within the bracket of equation (12) is negligible to our approximation. Hence we may put

$$\frac{a_0^2}{t} = \frac{2D(\rho_2-\rho_0)}{\sigma}$$

$$\sim 2\times10^{-6}\,\text{cm.}^2\text{sec.}^{-1} \text{ for water vapour in air.} \tag{13}$$

It will be noted that the vapour-density gradient at the surface of a drop varies only slightly as the drop grows; similarly, in the later stage of growth, when the drop is falling into undepleted gas, the density gradient will also not vary for this reason, contrary to

* See, for example, Herzfeld, *Kinetische Theorie der Wärme* (Vieweg, 1925), p. 340.

expectation, and no significant change in growth rate is to be expected between the initial stages of growth and the later stages when the drop is falling.

The equation of heat conduction, which determines the temperature of the growing drop, is of the same form as the diffusion equation and the approximation of equation (13) is also valid. Hazen (1942) has shown that in this approximation these two equations,

$$\frac{da_0^2}{dt} = \frac{2D}{\sigma}(\rho_2 - \rho_0),$$

$$\frac{da_0^2}{dt} = \frac{2\kappa}{\sigma\lambda}(\theta_2 - \theta_0),$$

where λ is the latent heat of condensation, κ the thermal conductivity and θ_2, θ_0 the temperatures far from the drop and at its surface, together with the relation between saturation vapour pressure and temperature, which may be written

$$p = A\exp(-b/\theta),$$

determine the rate of growth and the drop temperature after expansion to the volume ratio $(1+r)$ from the condition ρ_1, θ_1. He gives the expression

$$\frac{da_0^2}{dt} = \frac{2D\rho_1}{\sigma(1+r)}\left[1 - (1+r)^\gamma \exp\left\{-\frac{b}{\theta_1}\left(1 - (1+r)^{1-\gamma} - \left(\frac{\lambda\sigma}{2\kappa\theta_1}\right)\frac{da_0^2}{dt}\right)\right\}\right].$$
$$(14)$$

In Table IV we give the temperature of drops formed in nitrogen, hydrogen and helium at typical expansion ratios and for a final pressure about one atmosphere, while in Table V the values of a_0^2/t derived from equation (14) are shown for nitrogen and hydrogen (no value of the diffusion of alcohol vapour in helium is available). It is interesting to notice that condensation in the light gases is rapid, not only on account of the high diffusion coefficient but also because the growing drop remains cold.

TABLE IV. *Temperature of drops condensing from alcohol vapour* (°C.)

Permanent gas ...	N₂	H₂	He
Expansion ratio	1·16	1·15	1·10
$\theta_0 - \theta_2$	10°	4°	4°
$\theta_1 - \theta_2$	15·5°	15°	15°

Measurements of drop growth may be made by photographing a single drop using intermittent illumination, and by deducing the drop size from the rate of fall. This method has been used by Hazen (1942) and more recently by Barrett and Germain (1947); the results of both investigations are summarized in Table VI. The small variation of rate of growth with expansion ratio is not given here, and the values refer in all cases to a final gas pressure slightly greater than atmospheric. Hazen's measurements begin at about $t = 0.06$ sec. and those of Barrett and Germain at about $t = 0.15$ sec.; in both sets the zero of time is rather uncertain.

TABLE V. *Rate of drop growth* $(cm.^2sec.^{-1} \times 10^6)$ *from alcohol vapour in nitrogen, and hydrogen (equation* (14)). *Total final pressure about one atmosphere*

Permanent gas ...	N$_2$		H$_2$	
Expansion ratio	1·152	1·168	1·146	1·169
Rate of growth	5·2	5·65	27·5	31

TABLE VI. *Experimental values of drop growth* $(cm.^2sec.^{-1} \times 10^6)$ *in a final gas pressure of* 1·1–1·2 *atmosphere*

Workers	Permanent gas	Condensant		
		Water	25/75 water-alcohol	Alcohol
H	Hydrogen	—	—	18
	Helium	—	10	17
	Nitrogen	—	4·0	5·5
B and G	Air	7·5	4·4	—

There is excellent agreement between the experiments and the theoretical value of growth for nitrogen, but there is a discrepancy for hydrogen which can hardly be attributed to the approximation which is used by Hazen. It is more likely that the error lies in the experimental value which depends essentially on the application of Stokes's law to a falling 'hot' drop and in which no account is taken of the buoyancy of the heated sheath of gas surrounding the growing drop.

CHAPTER II

IONIZATION IN CLOUD TRACKS

2.1. Introduction

The ionization of an atom or molecule by a fast particle may leave the ejected electron with kinetic energy of any value between zero and a large fraction of the energy of the incident particle, the precise value of which depends upon the nature of the particle. For many purposes, the whole range of collisions may conveniently be divided into two categories, although the transition between the categories is not abrupt. Collisions at an impact parameter larger than the atomic dimensions occur frequently and give ionization with low kinetic energy; the collisions may be treated by the Williams-Weiszäcker analysis in terms of photoelectric absorption of the virtual equivalent radiation. Collisions, on the other hand, at impact parameters smaller than the atomic dimensions are rare and can simply be treated as between free particles, being described, in the appropriate system of co-ordinates, by the Rutherford scattering relation.

These categories correspond qualitatively to two forms of behaviour of practical importance. The 'photoelectrons' have as a rule so little kinetic energy that they are unable to produce further ionization, while the 'knock-on' electrons from elastic collisions are themselves able to produce 'secondary ionization'. A distinction is in practice made between 'primary ionization'—that produced by the direct interaction of the primary fast particle—and 'total ionization', which includes in addition ions produced indirectly by the faster secondary electrons from the primary process.

Finally, the more distant collisions lead to excitation as well as ionization of the atoms of the material traversed, and a small amount of ionization will occur at the absorption of the resultant radiation. This radiation will mainly be extremely soft, and although the part it plays in ionization by fast particles has not been studied in any detail, Hazen (1942) has concluded that the total number of ions formed at an appreciable distance from the main column of ionization is insignificant to the accuracy which can at present be reached in measurements of ionization.

The characteristic appearance of a cloud track, as it shows the distribution of ions along a particle trajectory, is well known. The ionization along the strongly scattered short trajectories of secondary electrons leads to blobs of close droplets which give to the track a beaded appearance. From the standpoint of trajectory measurements in the cloud chamber, apart from studies of ionization, the most important feature is the distribution of primary ionization. It is the primary ions which give the most precise trace of the trajectory of a fast particle, and for this purpose these, together with the smallest clusters, which before diffusion occupied a spread much smaller than the subsequent distribution under diffusion, are alone of value.

2.2. The energy transfer in ionizing collisions

The energy loss in ionizing collisions of particles passing through matter was treated classically by Bohr (1913, 1915), who gave a result of the form

$$-\frac{dE}{dx} = \frac{A}{\beta^2}(\ln k\beta^2 - \ln(1 - \beta^2) - \beta^2), \tag{15}$$

where $\beta = v/c$, A and k are constants given explicitly for a hydrogen-like atom, and where v is large compared with the orbital velocities of the electrons concerned. The behaviour of slow particles, for which this condition does not hold, is irrelevant for all cloud-chamber purposes.

The quantum-mechanical treatment is due to Bethe (1930), Bloch (1933), and others. Using the Thomas-Fermi model, Bloch gives the relation

$$-\frac{dE}{dx} = 2\pi N Z z^2 r_0^2 \frac{mc^2}{\beta^2}\left\{\ln\frac{mc^2 W}{(IZ)^2} + \ln\left(\frac{\beta^2}{1 - \beta^2}\right) + (1 - \beta^2)\right\}, \tag{16}$$

provided that $z/137\beta \ll 1$,

where a fast particle of charge ze and velocity βc traverses matter of N atoms/cm.³ of atomic number Z and of average ionization energy IZ, and where W is the maximum energy which the fast particle can transfer to a free electron. Here, I, determined from the loss of energy of α-particles in gold, is 13·5 V. The condition $z/137\beta \ll 1$ is met in all situations where ionization in the cloud chamber is of interest.

The result of equation (16) depends only upon the charge, ze, and the velocity βc, of the fast particle, except as regards the maximum single-energy transfer, W. For a fast particle of mass μ, and total energy $\gamma\mu c^2$, the value of W^* is

$$\frac{W}{(\gamma-1)\,\mu c^2} = \frac{2m\mu(\gamma+1)}{m^2+\mu^2+2m\mu\gamma}, \tag{17}$$

and at a given incident energy this quantity is clearly sensitive to the mass μ unless

$$\gamma \ll \mu/2m,$$

but since W appears only in the logarithmic term, the total effect is not great.

We have followed here the usual assumption that the average ionization energy of the relevant collisions can be written IZ, where I is a constant for all atomic numbers. The value $I = 13\cdot5$ eV., given by Bloch, is derived from the loss of energy of α-particles with a velocity of only 2×10^9 cm./sec. (comparable with the velocity of K electrons, $Z = 9$), but other results indicate that this is only very approximately of general application; R. R. Wilson (1941), for example, finds $I = 11\cdot5$ eV., from the stopping power of aluminium for 4 MeV. protons. In a recent investigation, Halpern and Hall (1948) have discussed the more exact treatment of average ionization energy in terms of the characteristic frequencies of the atoms of the material concerned, and show that appreciable specific effects are to be expected from particular materials. The values given by these workers and by the relation usually adopted are compared in Table VII.

TABLE VII. *Average ionization energies (eV.)*

Material	Lead	Iron	Water	Carbon	Air	Helium
$IZ = 13\cdot5Z$ (eV.)	1100	350	135	81	95	27
Halpern and Hall	1200	430	80	60	96	40

If it is assumed that the average energy loss per ion pair formed is independent of velocity, both for the primary particles and for

---

* For electrons, when the identity of the two particles after collision is uncertain, a modified relation obtains.

the secondary ionizing electrons, the number of ions formed per unit length of track will also be represented by an expression of the form of equation (16), whether we measure the total number of ions produced directly or indirectly, that is, the *total ionization*, or if we are able to measure the number of single ionizing events, the *primary ionization*.

The *primary ionization* may, but the *total ionization* cannot, be measured directly in the cloud chamber. As we have seen, the total ionization includes rare transfers in which a large fraction of the energy of the incident particle is passed to a secondary electron. Even if a statistically adequate sample were available, these very energetic secondaries would certainly pass out of the chamber and the ionization along them could not be counted. The quantity related to total ionization which can be measured in the cloud chamber is referred to as the *average* or *probable ionization*, and different workers have defined it in slightly different ways. The average ionization includes the ionization by secondary electrons up to a certain energy. This energy limit might be imposed directly by the apparatus, as a statistical cut-off in the neighbourhood of that energy transfer which will on the average happen once in the particular length of track available for measurement. It is more satisfactory, however, to set a lower limit than this, and to exclude clusters of ionization arising from secondaries of greater energy; not only is the statistical straggling of mean ionization from track to track reduced, but the difficulty of counting very large clusters of ionization is avoided. The *average ionization* should therefore be defined as being measured including secondary clusters of ions up to a certain specified size.

This definition of average ionization in effect replaces the large, mass-sensitive quantity W in equation (16) by a smaller quantity W' which is usually of the order 1000 eV. and which, for the range of primary energy which is of interest, is not mass-sensitive. Thus, average ionization, according to the Bloch relation, is a function only of particle charge and velocity.

2.3. Breadth of the column of ionization

For most purposes (§§ 6.1–6.18) we are able to regard the points of formation of ions as lying exactly on the trajectory followed by an ionizing particle. It was first pointed out by Williams, however, in 1931, that very energetic primary particles would produce ions at an appreciable distance from the particle trajectory, and that these ions were to be associated with the ultimate logarithmic increase of the Bloch expression with particle energy. The reason for this spread of the column of ionization can be understood by reference to Bohr's classical treatment.

In a collision of closest approach, p, by a fast particle of velocity $v(\beta \ll 1)$, the effective time of collision is of the order p/v and the energy transfer is proportional to the time integral of the interaction force, provided that $p/v < 1/\nu$ (where ν is the natural frequency of the electron concerned), and is negligible if $p/v > 1/\nu$. A limiting impact parameter, p', is thus fixed above which the collision is too slow for any effective transfer of energy. In the relativistic region, for the Lorentz system in which the electron is initially at rest, the collision time is reduced by a factor $(1 - \beta^2)^{-\frac{1}{2}}$, while the time integral of force is unchanged; then the limiting impact parameter, p', is increased by $(1 - \beta^2)^{-\frac{1}{2}}$, and, for extremely high energies, becomes large.

For an electron of energy 10^{12} eV., Williams showed that an appreciable number of ions would be formed up to 1 cm. from the trajectory (about 3 ions/cm. at distances between 0·5 and 1·0 cm. in oxygen at N.T.P.). This intrinsic broadening of ion tracks has not yet been observed, but it may just lie within the limits of observation for the most energetic cosmic-ray particles (see, however, § 2.6).

2.4. Measurements of track ionization

Primary ionization was measured for slow electrons by C. T. R. Wilson (1923), for β-particles by Williams and Terroux (1930) and by Skromstad and Loughbridge (1936), and for fast particles in the cosmic-ray region by Kunze (1933 a) and by Hazen (1943) and his co-workers. Williams and Terroux give the empirical expression $N_0 = 22\beta^{-1.1 \pm 0.2}$ ($22\beta^{-1.4}$ if the measurements of C. T. R. Wilson are also included) for the primary ionization N_0 ion pairs/cm. in air at N.T.P. for electrons in the velocity range $0.54 < \beta < 0.97$, while similar measurements by Skromstad and Loughbridge gave $N_0 = 19\beta^{-1.15}$, $0.89 < \beta < 0.98$. The large deviations of these results from the elementary variation with β^{-2} are notable. In the Bloch relation, the terms in β inside the bracket reduce the effective power of β by an amount which depends on the value of W' which is relevant, and Brode (1939) has concluded that these results are in accord with the Bloch relation. Hazen measured the primary ionization in helium at atmospheric pressure, saturated with a 25/75 water-alcohol mixture, for electrons near to the point of ionization minimum and for an unselected group of cosmic-ray mesons. The results were: N_0 (electron minimum) $= 7\cdot33 \pm 0\cdot12$ and N_0 (meson group) $= 7\cdot23 \pm 0\cdot13$. The latter figure is distinctly surprising, since the mean energy of the meson group would be expected to be high enough to lead to an appreciable contribution of ionization over and above the minimum, from the logarithmic increase. The apparent absence of this additional ionization is to be compared with the behaviour of the average ionization of mesons at sea-level and underground, also observed by Hazen, which is described below.

2.5. Average ionization

The average ionization has been measured for fast electrons and for cosmic-ray particles by Corson and Brode (1938), Locher (1937), Sen Gupta (1943), Hazen (1944), and others; the work of Hazen is probably the most critical yet undertaken. It is the average ionization, measured as a function of particle momentum, which is likely to play an increasingly important part in the interpretation of cloud-chamber data when particles of several different kinds are present, and the considerations which affect the precision and reliability of such counts are therefore of importance. It will be noticed that since average ionization is a function of velocity alone (we will assume, in accordance with experience, that the charge of the particle is not in question), in which each value of ionization corresponds to two values of velocity, known values of momentum and ionization correspond in principle to two alternative values of mass only. However, in practice, if we exclude the possibility of charged particles of less than electronic mass, the ionization-momentum diagram (fig. 4) divides into two sections, A, above the electron line, in which the ionization-momentum relation corresponds sensitively to a single-mass value, and B, below the electron line, in which points correspond insensitively to two alternative masses. Region A is of the greatest general importance, region B so long as a small number of particles only is known, and provided the logarithmic increase of ionization with energy is satisfactorily established experimentally, may prove useful for purposes of identification at high momenta (see, however, §2.6 below).

Some causes of uncertainty in a drop count are immediately apparent. It is difficult, particularly when mixed condensing vapours are used, to state the exact composition of gas in which an ion count is made; to permit counting, tracks are allowed to diffuse

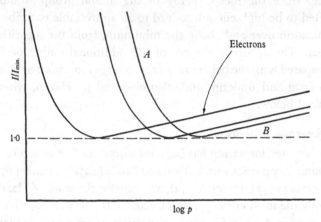

Fig. 4. Ionization density of singly charged particles as a function of momentum.

for a fraction of a second before being fixed by expansion, and a certain amount of recombination is possible; soft X-rays from excited atoms may lead to ion formation at an appreciable distance from the geometrical trajectory. Hazen, however, has reached the conclusion that the greatest uncertainty lies in the degree of condensation efficiency which is reached. We have already pointed out that the drop nucleus for condensation on an ion is subject to statistical fluctuations which, unlike those on uncharged aggregates, are normally of little concern. Because the ion limits lie close to the onset conditions for background condensation, it is not possible to ensure 100 % condensation on ions merely by allowing an ample margin of expansion ratio, for this leads immediately to an unacceptably dense background. Hazen draws positive and negative ions from a single track into completely separate columns, and adopts as a criterion of condensation efficiency the ratio of number of drops in the positive- and negative-ion columns. With approximately 75/25 ethyl alcohol-water mixture, condensation on more than 95 % of positive ions occurs when the ratio of drops in positive and negative columns is as 10 is to 1, and on substantially 100 % of positive ions when this ratio is less than 5 to 1. For a very large range of vapour compositions, the positive-ion condensation is 100 % when the same ratio is less than 2 to 1.

The average ionization of electrons has been investigated by Corson and Brode in air and by Sen Gupta in argon; their results are summarized in Table VIII.

TABLE VIII. *Average ionization by electrons in air at* N.T.P. (*ion pairs/cm.*)

Electron momentum (eV./c.)	Corson and Brode (air)	Sen Gupta (argon)
5×10^5	78 ± 6	—
1×10^6	56 ± 1	—
2×10^6	54 ± 3	—
3×10^6	—	$38 \cdot 0 \pm 0 \cdot 7$
4×10^6	48 ± 1	—
7×10^6	56 ± 1	$42 \cdot 5 \pm 0 \cdot 6$
$1 \cdot 5 \times 10^7$	59 ± 2	—
3×10^7	61 ± 2	$48 \cdot 5 \pm 0 \cdot 6$
7×10^7	—	$54 \cdot 5 \pm 0 \cdot 8$
2×10^8	—	$56 \cdot 5 \pm 1 \cdot 5$
5×10^8	—	$63 \cdot 0 \pm 2 \cdot 1$

In this table, Sen Gupta's value of W' is about 1300 eV.; the corresponding value for Corson and Brode's results is not stated but is certainly higher. The rise of ionization at high momenta seems well established, although the results are discrepant, since the average ionization in argon is expected to exceed that in air.

Similar measurements for mesons have been made by Sen Gupta and Hazen. The former measured ionization for particles of momenta known to an accuracy determined by the maximum detectable momentum 3×10^9 eV./c. (§§ 6.10, 6.11), while Hazen measured unselected groups of mesons at sea-level and 100 ft. underground, which we tabulate (Table IX) at estimated mean momenta 2×10^9 eV./c. and 2×10^{10} eV./c. Sen Gupta includes secondary clusters up to 40 and Hazen up to 25 ions of either sign (W' respectively about 1300 and 800 eV.). For mesons, the evidence of an increase of average ionization in the relativistic region is not so clear; the evidence lies solely in the measurement at about 4×10^9 eV./c., and it is difficult to give this value more weight than that of Hazen at 2×10^{10} eV./c. The standard deviations quoted by Sen Gupta (Tables VIII and IX) are underestimated by a factor of about two, since he assumes all ion pairs, and not only primary ionizations, to be independent. Again, the ionization in argon appears low relative to that in air.

TABLE IX. *Average ionization by mesons at* N.T.P. *(ion pairs/cm.)*

Meson momentum (eV./c.)	Sen Gupta (argon)	Hazen (air)
3×10^8	$41 \cdot 5 \pm 0 \cdot 6$	—
8×10^8	$41 \cdot 0 \pm 0 \cdot 6$	—
2×10^9	$42 \cdot 5 \pm 0 \cdot 8$	$51 \cdot 0 \pm 1 \cdot 7$
4×10^9	$47 \cdot 0 \pm 0 \cdot 5$	—
2×10^{10}	—	$50 \cdot 0 \pm 2 \cdot 0$

The very general nature of the considerations upon which the logarithmic increase of ionization is based make it most probable that the increase of ionization does in fact occur for mesons, but the figures in Table IX serve to emphasize the difficulty at present in using average ionization as a property for identification of particles at high momenta (see also §2.6).

Hazen also surveyed the frequency of occurrence of secondary clusters rather larger than his average ionization limit, from 25 drops of either sign (~ 800 eV.) to about 310 drops (10,000 eV.). The total amount of ionization in this range of secondaries was roughly 8 % of that excluding all secondaries of greater than 800 eV., and one secondary in this energy range occurred, on the average, in each 15 cm. of track. Hazen compared the size distribution of these secondary clusters with Rutherford scattering of free electrons and of electrons in the K shell of oxygen or nitrogen;

$$N(E)\,dE = k\frac{dE}{E^2} \text{ cm.}^{-1}\text{eV.}^{-1}, \qquad (18)$$

where k is 98 for free electrons and 71 for the K electrons. He considered that the actual situation could be represented by an intermediate relation, and the agreement of the measured distribution with this assumption is excellent.

2.6. Polarization effects at high momenta

The treatments of the process of ionization by fast particles considered in earlier sections deal essentially with the effect on an isolated atom of the passage of the ionizing particle. Fermi (1940), and later, Halpern and Hall (1940, 1948), have considered the extent to which the assumption of isolation is permissible in actual matter.

It is clear that very close collisions will not be affected by the presence, at much greater distances, of other atoms. Fermi, therefore, treats separately the energy transfer from a fast particle to distances greater than b from its trajectory, where b is some distance rather greater than the atomic dimensions, and derives the classical value of energy transfer to greater distances considering only electrons of the matter traversed which have a single resonant frequency. (The extension to a series of resonant frequencies has been given by Halpern and Hall.) The classical transfer is the Poynting flux across a cylinder of radius b surrounding the particle trajectory, and the way in which this is modified in the presence of the polarizable electrons of the medium is found to depend essentially upon whether the phase velocity of the radiation field of the ionizing particle is greater or less than its particle velocity. Writing ϵ for the dielectric constant of the medium traversed, the classical rate of energy loss at distances greater than b ignoring polarization of the medium, is (notation of equation (16) above)

$$\left(-\frac{dE}{dx}\right)_{>b} = 2\pi N Z z^2 r_0^2 \frac{mc^2}{\beta^2}\left(\ln\frac{mc^2\beta^2}{3\cdot17\pi N Z e^2 b^2} + \ln\frac{(\epsilon-1)\beta^2}{1-\beta^2} - \beta^2\right),$$
(19)

and the reduction in this quantity when polarization effects are included is

$$2\pi N Z z^2 r_0^2 \frac{mc^2}{\beta^2}(\ln\epsilon) \quad (\beta < \epsilon^{-\frac{1}{2}})$$
(20)

and

$$2\pi N Z z^2 r_0^2 \frac{mc^2}{\beta^2}\left(\ln\left(\frac{\epsilon-1}{1-\beta^2}\right) - \frac{1-\epsilon\beta^2}{\epsilon-1}\right) \quad (\beta > \epsilon^{-\frac{1}{2}}).$$
(21)

When more than one electron frequency exists in the medium

traversed, the effects are substantially similar, although it is no longer possible to write the above relations in terms of a single dielectric constant, nor is the elimination of the logarithmic increase of energy loss at high energies ($\beta > \epsilon^{-\frac{1}{2}}$) complete.

In gases, and for slow particles ($\beta < \epsilon^{-\frac{1}{2}}$), the modification of energy transfer beyond a distance b is insignificant, but for faster particles ($\beta > \epsilon^{-\frac{1}{2}}$) the change is considerable, and in particular the logarithmic increase of ionization would practically cease (fig. 5).

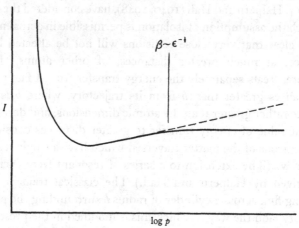

Fig. 5. Schematic diagram of the modification of the ionization-momentum relation by polarization effects. The full curve shows the result given by Fermi (1940); the more complete treatment by Halpern and Hall (1948) leads to a curve ($\beta > \epsilon^{-\frac{1}{2}}$) intermediate between the full and broken lines.

In real dielectrics we may take to order of magnitude the condition $\beta = \epsilon^{-\frac{1}{2}}$ to give the point of transition between equations (20) and (21). Thus the momentum of the transition point is

$$pc = \mu c^2 \frac{\beta}{(1 - \beta^2)^{\frac{1}{2}}} = \mu c^2 (\epsilon - 1)^{-\frac{1}{2}}. \tag{22}$$

In the usual mixtures, the condensible vapour will make a considerable contribution to the dielectric constant of the chamber gas, and this will probably lie in the range

$$2 \times 10^{-4} < (\epsilon - 1) < 6 \times 10^{-4}.$$

Thus
$$70 > \frac{pc}{\mu c^2} > 40,$$

and the transition-point momentum for electrons is $p \sim 3 \times 10^7$ eV./c. and for mesons ($\mu = 200$) $p \sim 6 \times 10^9$ eV./c. At greater momenta little or no increase in ionization density would be expected.

Existing experimental evidence does not allow a definite decision to be taken as to the correctness of this result. As regards the average ionization by electrons, the measurements of Corson and Brode extend only to the probable critical momentum, but those of Sen Gupta seem definitely to favour a continuation of the logarithmic increase of ionization to considerably higher momenta. For mesons, on the other hand, Sen Gupta's results do not reach as far as the critical momentum, but the measurements by Hazen which have already been quoted, in which the ionization of mesons at sea-level and 100 ft. underground were compared, suggest that the logarithmic increase of ionization does not extend to much higher momenta than about 2×10^9 eV./c. This conclusion, of course, depends essentially upon the assumption* that the meson beams used by Hazen are adequately described by the mean momenta used above (Table IX). The close equality, also observed by Hazen, between the primary ionization of electrons near to the ionization minimum and of an unrestricted meson group is not relevant, since the average momentum of such a group of mesons is certainly no higher than about 2×10^9 eV./c. In the face of this rather confusing situation, the whole of the ionization curve above the minimum value must at present be regarded as too uncertain to be of any value at all for the purposes of particle identification.

2.7. Extension of the Fermi treatment.

The analysis developed by Fermi can in principle be used over a cylindrical surface of any value of the radius b, although, when b becomes large, approximations which are valid for b of the order of atomic dimensions can no longer be used. A particular case of importance is that in which b becomes infinite; then the net flux of energy out of the cylinder represents a free emission of radiation. This is the Cerenkov radiation (1937), and Fermi shows that this essentially classical radiation effect is given in the correct form in his treatment.

* The figures are the author's and are not quoted from Hazen's paper.

CHAPTER III

THE PRODUCTION AND MAINTENANCE OF SUPERSATURATION

The operation of all cloud chambers which have been at all extensively applied depends upon the production and persistence for a certain limited time of the critical supersaturation necessary for track condensation. We shall discuss here the conditions of development of supersaturation, and the factors which control its duration, finally referring briefly to attempts which have been made to produce apparatus in which a gas volume is maintained continuously at the level of critical supersaturation.

3.1. Elementary features of a volume-defined expansion

In the classical apparatus of C. T. R. Wilson, supersaturation is developed by adiabatic expansion in which a light piston is moved in a geometrically defined manner, that is, in a *volume-defined* expansion. The amount of expansion is described by v_2/v_1, the ratio of the final volume of the chamber space to the initial volume (fig. 6), the so-called expansion ratio. For air saturated with water vapour, the expansion ratio to attain fairly complete condensation on ions with little background condensation in a clean chamber is about $v_2/v_1 = 1\cdot30$, and it will be useful to give a numerical example of the effect of such an expansion. If the initial temperature, θ_1, is $10°$ C., the partial pressure of water vapour will be $9\cdot2$ mm. Hg, and the ratio of the principal specific heats for the mixed gas will be very little different from that for dry air, say $\gamma = 1\cdot40$. Thus the final temperature, θ_2, of the gas will be given by

$$1\cdot30 = \left(\frac{v_2}{v_1}\right)_{\text{air}} = \left(\frac{\theta_1}{\theta_2}\right)^{1/(\gamma-1)} = \left(\frac{283}{\theta_2}\right)^{2\cdot5},$$

or
$$\theta_2 = -18\cdot2°\,\text{C.},$$

while the final pressure, p_2, of water vapour will be

$$p_2 = 9\cdot2\,\frac{\theta_2}{1\cdot30\theta_1}\,\text{mm. Hg}$$

$$= 6\cdot4\,\text{mm. Hg.}$$

The saturation vapour pressure, p_{20}, at $\theta_2 = -18\cdot2^\circ$ C., however, is much smaller;
$$p_{20} = 1\cdot0 \text{ mm. Hg,}$$

and so when the expansion is completed, the cooled gas is super-saturated about sixfold. The condensed droplets which are formed are liquid, not solid, for they are the seat of liberation of the large latent heat of condensation; they are, in fact, for the example given, never very far below the bulk freezing-point (§ 1.8 above).

The conditions under which ice is sublimed from supersaturated vapour have recently been investigated by Cwilong (1947), who, working with water vapour, used the property that freezing will be induced in supercooled water at the bottom of the expansion

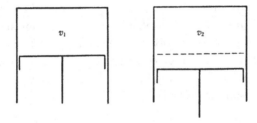

Fig. 6. Volume-defined cloud chamber.

chamber if any of the condensed aggregates falling into it after expansion are ice rather than liquid water. He found that no ice particles are formed at adiabatic expansion provided that the final temperature is not lower than $-41\cdot2^\circ$ C., and that until this limit is reached, no condensation takes place in clean gas unless an expansion of at least $1\cdot25$ is made. However, if adiabatic expansion leads to a final temperature below $-41\cdot2^\circ$ C., ice crystals are formed, and these are first noticed at this final temperature whether or not the adiabatic expansion leading to it was as great as $v_2/v_1 = 1\cdot25$. The formation of ice crystals was noted for expansions as small as $v_2/v_1 = 1\cdot07$, and Cwilong considers this to be a lower limit of expansion ratio set rather by the difficulties of observation than by the essentials of the phenomenon.

The nature of the condensation nuclei upon which ice is formed at small expansion ratios below $-41\cdot2^\circ$ C. is not well established, but there is some indication that these may be ions.

Referring again to the normal temperature range, in a monatomic gas, for example in argon, an expansion leading to the same temperature drop as in air:

$$(v_2/v_1)_{\text{air}}^{1\cdot40-1} = (\theta_1/\theta_2) = (v_2/v_1)_{\text{argon}}^{1\cdot67-1},$$

gives a supersaturation greater in the ratio $(v_2/v_1)_{\text{air}}/(v_2/v_1)_{\text{argon}}$, about $1\cdot11$: thus an expansion ratio $1\cdot17$ in argon leads to greater supersaturation than expansion ratio $1\cdot30$ in air. The lower expansion ratio which is necessary in the monatomic gases is of general advantage, since the piston movement for which mechanical provision is made is reduced. The full advantage, however, is only attained when the partial pressure of the polyatomic vapour is small compared with that of the permanent gas. For water vapour ($p_1 = 9\cdot0$ mm. at $10°$ C., $\gamma = 1\cdot30$), this condition is met for a total pressure about atmospheric, but for alcohol and for the widely used water-alcohol mixtures (alcohol: $p_1 = 24$ mm. at $10°$ C., $\gamma = 1\cdot13$) it is not, and the necessary expansion ratio is noticeably reduced at pressures of several atmospheres as compared with the value at atmospheric pressure.

3.2. Speed of expansion

For most purposes there is little difficulty in attaining a speed of motion of the chamber piston sufficient to make the expansion very closely adiabatic. The slowest tolerable speed is closely connected with the time persistence of supersaturation after it has once been established, and the values which are quoted in a later section (§ 3.6) for the sensitive time of a chamber may be taken to order of magnitude as the time of expansion within which the supersaturation developed is significantly that for an instantaneous expansion. The discussion of sensitive time makes clear the physical mechanism operating to modify the growth of supersaturation.

It was first shown by Powell in 1928 that when the partial pressure of condensible vapour becomes a large fraction of the whole pressure, other factors operate which greatly restrict the growth of supersaturation. Even with very rapid piston motion, the supersaturation which is reached is then much smaller than that deduced from adiabatic expansion to the particular volume expansion ratio used. This effect is attributed to evaporation of 'hot' vapour from the free liquid surface in the chamber (necessary to ensure complete pre-expansion saturation), which is not cooled at the expansion. Powell has estimated that following an instantaneous expansion, $v_2/v_1 = 2 \cdot 0$, for pure steam at $100°$ C., the time for the original steam pressure to be restored by evaporation would be of the order 10^{-4} sec.; a cloud chamber working in this condition would require a mechanical expansion completed in a time short compared with this figure. Under more normal conditions, Joliot (1934) has shown that from an initial temperature of $17°$ C., the working expansion ratio of a chamber operating at substantially constant speed (the speed is not quoted but the expansion is certainly 'fast'), increases from $1 \cdot 305$ at atmospheric pressure to about 2 when only saturated water vapour remains. The difference represents a departure from adiabatic conditions during the time of the expansion, and should be compared with the variations of sensitive time of the same chamber under similar conditions (§ 3.6 below).

It is apparent that the operation of a conventional cloud chamber with pure vapours is a matter of considerable difficulty, and that the permanent gas in the chamber plays a quite essential role in its operation by restricting evaporation from free liquid in the chamber. This in the first place allows the necessary characteristic supersaturation for track condensation to be attained at a reproducible expansion ratio, but it assists further, tending to maintain supersaturation and so allowing droplets to continue to grow right up to the moment of photography.

3.3. The critical supersaturation limits as a function of temperature

We have already found (§§ 1.3, 1.4) on theoretical grounds that the temperature variations of the ion limit of supersaturation and of the background limit are expected to be of the form

$$\left. \begin{aligned} \ln\left(\frac{p}{p_0}\right)_{\text{crit.}} &\sim T\theta_2^{-1} \quad \text{(ion limit),} \\ \ln\left(\frac{p}{p_0}\right)_{\text{crit.}} &\sim T^{\frac{3}{2}}\theta_2^{-\frac{3}{2}} \quad \text{(background limit).} \end{aligned} \right\} \tag{23}$$

Thus the limits for ion and background condensation are reduced

3

with increasing temperature, while at the same time the relative separation between them becomes less. These properties are qualitatively confirmed in Powell's experiments, although the numerical values which he gives show a more rapid variation. (Powell's values are all, however, extrapolated measurements in which vapour evaporation during expansion was appreciable.) Since the separation in supersaturation between the optimum of ion condensation and the development of serious background is not large, the variation of the separation with temperature is probably of practical significance, making it desirable to keep the chamber temperature at the lowest convenient level.

It is interesting to note the rather small range of temperatures through which a cloud chamber using water vapour can readily be operated. Cwilong has established an effective lower limit of the final temperature at about -40° C., and so of the initial temperature for ion condensation of about -10° C. The upper limit is not so clearly defined, but operation is becoming difficult when the initial temperature is about 50° C. It is perhaps fortunate that the applications of the cloud chamber do not as a rule impose any restriction on the steady temperature of the chamber itself.

3.4. Volume-defined and pressure-defined chambers

The cloud-chamber process is characterized by the adiabatic expansion of saturated gas leading to a reduction of temperature from, say, θ_1 to θ_2, and, for a particular initial state, the degree of expansion for a specific condensation condition is defined by θ_1/θ_2. We cannot conveniently in practice measure the expansion with direct reference to the reduction of temperature in the gas; what is effectively controlled is either the change of volume or the change of pressure, and these alternatives correspond to *volume-defined* and *pressure-defined* operation. The classical cloud chamber constructed by C. T. R. Wilson in 1912 was a volume-defined chamber; the prototype of the pressure-defined chamber, also due to Wilson, was described in 1933.

It is clear that α and β, the volume and pressure expansion ratios corresponding to a temperature reduction θ_1/θ_2 in adiabatic expansion, are given by

$$\alpha = \frac{v_2}{v_1} = \left(\frac{\theta_1}{\theta_2}\right)^{1/(\gamma-1)},$$
$$\beta = \frac{p_1}{p_2} = \left(\frac{\theta_1}{\theta_2}\right)^{\gamma/(\gamma-1)},$$

(24)

and that, therefore, $\qquad \alpha^{\gamma} = \beta.$

The magnitude of a volume-defined expansion is controlled by direct mechanical means and requires no comment here. A pressure change of fixed magnitude is rather more complicated: basically, a pressure difference is established between two vessels, one, that at higher pressure, being the cloud chamber, and the expansion is made by connecting the vessels and allowing pressure between them to equalize. The analysis of this situation was first given by Herzog (1935).

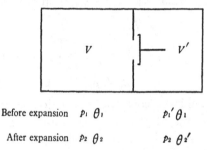

| Before expansion | $p_1 \; \theta_1$ | $p_1' \theta_1$ |
| After expansion | $p_2 \; \theta_2$ | $p_2 \; \theta_2'$ |

Fig. 7. Basic features of a pressure-defined chamber.

Let the vessels V, V' (fig. 7) be initially at temperature θ_1, and respectively at pressures p_1 and $p_1'(p_1 > p_1')$. The vessels are connected and come instantaneously to pressure p_2 and temperatures θ_2, θ_2'. The condensation condition in the first vessel is thus described by the temperature ratio θ_1/θ_2. The total mass of gas in the system and its total energy are unchanged, so

$$\frac{1}{\theta_1}(p_1 V + p_1' V') = p_2\left(\frac{V}{\theta_2} + \frac{V'}{\theta_2'}\right),$$

$$p_1 V + p_1' V' = p_2(V + V'),$$

and, in particular,

$$\beta = \frac{p_1}{p_2} = \frac{V/V' + 1}{V/V' + p_1'/p_1}.$$

(25)

3-2

Two conditions are of special practical importance:

$$\frac{V}{V'} = 0, \quad \beta = \frac{p_1}{p_1'} \quad \text{(expansion into an atmosphere)},$$

$$\frac{p_1'}{p_1} \to 0, \quad \beta \to 1 + \frac{V'}{V} \quad \text{(expansion into a vacuum)}.$$

C. T. R. Wilson's apparatus of 1933 is of the first type, and is technically extremely straightforward. It suffers from the disadvantage, however, that the expansion ratio is sensitive to changes of the atmospheric pressure, p_1', in the absence of a complex control to maintain $p_1 = \beta p_1'$. The ratio, β, at expansion into a vacuum, on the other hand, is substantially independent of p_1, and may be shown also to depend on p_1', the residual pressure in the vacuum chamber, on a very open scale which offers an excellent means of controlling the expansion ratio. If we write, in such a chamber,

$$p_1'/p_1 = \epsilon \ll 1,$$

then
$$\beta = \frac{V/V' + 1}{V/V'(1 + \epsilon V'/V)} = \left(1 + \frac{V'}{V}\right) - \epsilon \frac{V'}{V}\left(1 + \frac{V'}{V}\right) + \dots \epsilon^2 \dots$$
$$= \beta_0 - \epsilon \beta_0 (\beta_0 - 1), \tag{26}$$

where β_0 corresponds to complete evacuation of the receiving vessel. As an example, if $\beta_0 = 1 \cdot 44$ (corresponding to $\alpha = 1 \cdot 30$, the ion-condensation optimum for water vapour in air), and if $p_1 = 76$ cm. Hg, then

$$\beta = 1 \cdot 440 - 0 \cdot 0083 p_1',$$

where p_1' is also in cm. Hg. The control of expansion ratio is then shown in Table X, which also shows the variation of expansion ratio with changes of initial chamber pressure.

TABLE X. *Expansion β from a volume V at 70, 76 and 82 cm. Hg into a volume $0 \cdot 440V$ at p_1' cm. Hg*

(that is, over a range of initial pressures for which the composition of the gas is not varied significantly as regards its effective γ.)

p_1	p_1'				
	0	1	2	3	4
70	1·440	1·432	1·425	1·417	1·409
76	1·440	1·432	1·423	1·415	1·407
82	1·440	1·431	1·422	1·413	1·404

3.5. Expansion characteristics in volume-defined and pressure-defined chambers

The essential difference in the development of the expansion in volume-defined and pressure-defined chambers is shown schematically in fig. 8, in which, for ease of comparison, the pressure changes inside the chamber are illustrated. In a volume-defined chamber, the piston will, as a rule, be accelerated for the greater part of its motion, the acceleration remaining constant or even increasing with time, but because the mass of the piston is appreciable, it starts to move relatively slowly. The motion, on the whole, is likely to be completed in shorter time than in a pressure-defined apparatus.

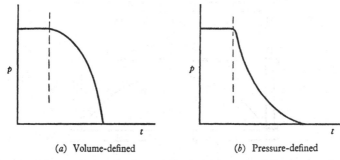

(a) Volume-defined   (b) Pressure-defined

Fig. 8. Characteristics of volume-defined and pressure-defined expansions.

Here the expansion starts quickly, since the total mass to be set in motion is very small; for the same reason, however, the inertia of rapidly moving gas at the exit hole is small and the rate of efflux falls off with decreasing driving pressure difference. If the motion were appreciably underdamped, puffs of returned gas might be troublesome while the persistence of supersaturation would clearly be short. In practice (see §3.6 below), the pressure-defined chamber consists of more than two coupled vessels and the detailed motion is complicated although the general features are as we have indicated.

Both of the modes of expansion shown in fig. 8 are of value, the volume-defined expansion being of particular importance when speed of expansion is to be stressed, while the pressure-defined expansion will be shown to lead more readily to a prolonged persistence of supersaturation.

3.6. Use of a porous diaphragm

The pressure-defined cloud chamber described by C. T. R. Wilson in 1933 is shown diagrammatically in fig. 9, where the whole space above the bung, B, corresponds to the left-hand, high-pressure vessel in fig. 7. This volume is divided (i) by a thin rubber membrane, R, which merely serves to isolate the working gas from the atmosphere, and may be dispensed with if a new supply of suitable gas is available for each expansion, and (ii) by the gauze diaphragm, G. The primary purpose of this diaphragm is to ensure non-turbulent motion in the upper section, C, of the chamber; it

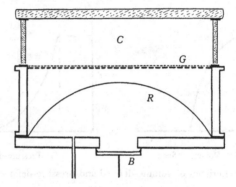

Fig. 9. Pressure-defined cloud chamber (C. T. R. Wilson, 1933).

is necessary for this purpose in all pressure-defined chambers constructed for rapid expansion, and usually consists of a gauze or perforated metal sheet covered with tightly stretched thin black ('ring') velvet, thus forming a particularly good photographic background. It is found that great freedom of design is possible as regards the whole space below the gauze.

An exactly similar diaphragm is now very frequently introduced into a chamber of the conventional volume-defined type, where it permits relaxations of piston design which are particularly valuable in the vertically operated chambers used in cosmic-ray investigations. It is clear that the introduction of the diaphragm brings such a chamber into an intermediate category between the two basic types which have been discussed.

3.7. The persistence of supersaturation

The supersaturation arising from adiabatic expansion is an unstable condition, since the wall of the containing vessel remains at a higher temperature than the enclosed gas. Accordingly, as soon as supersaturation has been established, heat exchange between the expanded gas and the chamber walls operates to destroy the supersaturation. If the effect is appreciable while the expansion is still being made, the expansion will not be strictly adiabatic, and a limiting speed, dependent upon the size and geometry of the chamber and the nature and pressure of the filling gas, may be stated below which the departure from adiabatic conditions will be appreciable. When expansion has been completed, the supersaturation immediately begins to fall, in part because of the heat exchange with the chamber walls already noticed, but also, when condensation is taking place, because of the liberation at condensed droplets of the heat of condensation. Moreover, under particular conditions (§3.2) evaporation from free liquid in the chamber may tend also to reduce supersaturation. The persistence of supersaturation is of particular importance when it determines the collecting time over which the chamber is open to receive particles occurring at random in time. For this purpose the chamber conditions which yield a long duration of supersaturation are exploited, and, further, modifications of the operating scheme of the chamber may be used to prolong artificially this collecting time. It is important to notice, however, that the sensitive time of a cloud chamber, whether simple or artificially prolonged, is not a distortion-free time; the factors which distort cloud tracks are operative at latest from the act of expansion, and precise geometrical measurements in any photograph can only be made on tracks much younger even than the normal sensitive time of a chamber of moderate size.

Persistence of supersaturation in a volume-defined chamber, ignoring condensation

If we ignore the effect of any condensation which takes place, supersaturation after the expansion is completed is reduced by conduction into the gas of heat from the chamber walls which were, of course, not cooled at the adiabatic expansion of the gas. The following analysis, due to E. J. Williams (1939a), shows that two domains of the chamber are concerned, a thin layer inside the walls which is heated directly by conduction from the walls, and which therefore expands, and the main central volume of the chamber, into which the conduction of heat does not penetrate appreciably. This main volume is thus heated only by compression, and the supersaturation in it, while decreasing with time, remains uniform and higher than that of the surface layer. Over the usable volume of the chamber, therefore, supersaturation sufficient for track condensation is lost at the same instant. The interval from the establishment of full supersaturation to this instant has been called the 'sensitive time' of the chamber.

Let $(1+r)$ be the minimum expansion ratio for track condensation, and let $(1+r+\delta r)$ be the ratio initially established. Now if θ_1 is the gas temperature before expansion, θ_2 the corresponding temperature after adiabatic expansion to the ratio $(1+r+\delta r)$ and $\theta_2+\delta\theta$ that after expansion to the ratio $(1+r)$, then

$$\frac{\delta\theta}{\theta_1-\theta_2} = \frac{\delta r}{r}, \tag{27}$$

where, in practice, $\delta r/r$ may be of the order 0·01.

We consider the layer of gas heated from the chamber wall, and write down the temperature distribution through this layer at time t ignoring, at this point, the volume change of the layer. The temperature, $\theta_2+\psi(\theta_1-\theta_2)$, at distance x from the solid wall is then given by

$$\psi = 1 - \frac{2}{\sqrt{\pi}}\int_0^{x/\sqrt{(4a^2t)}} e^{-z^2}\,dz, \tag{28}$$

where $a^2 = \kappa/\rho s$, κ is the thermal conductivity of the gas, ρ its density and s its specific heat at constant pressure.

The total *increase* of volume, δV, of the heated boundary layers over S, the whole surface area of the chamber wall, up to time t is therefore

$$\delta V = S\frac{\theta_1-\theta_2}{\theta_2}\int_0^{\infty}\psi\,dx = 1\cdot14S\frac{\theta_1-\theta_2}{\theta_2}a\sqrt{t}, \tag{29}$$

and this must also be the *reduction* in volume of the main body of gas away from the walls. The temperature of the main gas volume is thus increased by

$$\delta\theta = (\gamma-1)\frac{\delta V}{V}\theta_2, \tag{30}$$

where V is the total volume of the chamber.

If we now put for $\delta\theta$ the variation of temperature defined in equation (27), and for δV the value of equation (29), we find for τ_0, the sensitive time of the chamber,

$$\tau_0 = 0.77 \frac{\rho s}{\kappa (\gamma - 1)^2} \left(\frac{V}{S} \right)^2 \left(\frac{\delta r}{r} \right)^2. \tag{31}$$

Substitution of this value of τ_0 in equation (28) shows immediately that during the sensitive period of a chamber of normal dimensions and pressure, conductive heating from the chamber walls is only appreciable for a surface layer of the order of 10 % of the linear dimensions from the chamber wall. The whole useful volume of the chamber thus lies in the compressed central mass of gas, and throughout it sensitivity for track condensation ceases at the same instant.

The effect of condensation

If no condensation were to take place, the sensitive time would increase as $(\delta r)^2$. Each condensing drop, however, plays a similar part to the chamber wall, being a centre from which heat is conducted out into the surrounding gas, while, in addition, it is the centre of a region which has been depleted of vapour. With increasing supersaturation the background condensation increases rapidly, and a point is reached when the background droplets exert greater control on the supersaturation throughout the chamber than do the wall phenomena, and hence determine the sensitive time. In this condition, the sensitive time is rapidly reduced as the expansion ratio is still further increased, and there is thus an optimum expansion ratio which for any chamber leads to the greatest sensitive time.

A treatment of the control of sensitive time by condensation has been given by Hazen (1942) in terms of n, the density of condensation (drops/cm.³), and da_0^2/dt, the rate of drop growth (§ 1.8). The heating effect of condensation yields a result strictly comparable with that given by Williams for heating from the chamber wall,

$$\tau_1^{\frac{3}{2}} = \frac{1}{4\pi} \frac{\rho s}{\kappa(\gamma - 1)} \frac{1}{\frac{2}{3}n(da_0^2/dt)^{\frac{1}{2}}} \left(\frac{\delta r}{r} \right), \tag{32}$$

while the effect of depletion of the available vapour may be considered by equating m, the amount of vapour condensed per cm.³ up to time τ_2,

$$m = \tfrac{4}{3}\pi \sigma n \left(\frac{da_0^2}{dt} \right)^{\frac{3}{2}} \tau_2^{\frac{3}{2}}$$

with the total vapour available, M g./cm.[3] above that required for critical supersaturation; thus

$$\tau_2^{\frac{3}{2}} = \frac{3}{4} \frac{M}{\pi \sigma n} \left(\frac{da_0^2}{dt} \right)^{-\frac{3}{2}}. \tag{33}$$

Determined in this way, τ_1 and τ_2 are comparable in magnitude, but the nature of their action is very different. Only a slight depletion of vapour is necessary to suppress further new condensation, and τ_2 is given on the assumption that this slight depletion is attained uniformly throughout the gas. In fact, a very much greater depletion takes place close to the drop and a correspondingly smaller change elsewhere, and the concept of a definite time at which new condensation will cease throughout the chamber volume cannot properly be applied to this process. It is likely that the effective sensitive time is somewhat less than would be given by τ_0 and τ_1 only, but the difference is probably not great.

The method of combining the effect of heating by condensation with that from the chamber walls is straightforward, since both contribute in the same way to the compression of the main bulk of the gas. Hazen shows that $\bar{\tau}$, the resultant sensitive time of the chamber, is connected with τ_0 and τ_1 by the expression

$$\left(\frac{\bar{\tau}}{\tau_1} \right)^{\frac{3}{2}} + \left(\frac{\bar{\tau}}{\tau_0} \right)^{\frac{1}{2}} = 1. \tag{34}$$

The values of $\bar{\tau}$ corresponding to a measured development of background condensation (lower part of diagram) are shown in fig. 10, for 25/75 water-alcohol condensing in nitrogen, together with Hazen's experimental points in a chamber 30 cm. diameter. The best performance is reached in this example for a density of background condensation rather less than 10 drops/cm.[3]

It must be noticed that for larger chambers the transfer of control will take place at an even lower level of background condensation. Thus in large chambers ($d > 30$ cm.) it will be unusual for the sensitive time to be controlled by the size-sensitive wall heating, and the practical sensitive time will as a rule be independent of size beyond dimensions of this order. Similar considerations will govern the sensitive time of a high-pressure chamber, in which (see footnote, p. 46), after a certain period of collection, sensitivity will be terminated by the action of the total number of ions formed by particles entering during this time.

Measurements of sensitive time may be made by various methods. Williams used a source of known strength, for which the mean number of tracks was proportional to the sensitive time, and a similar

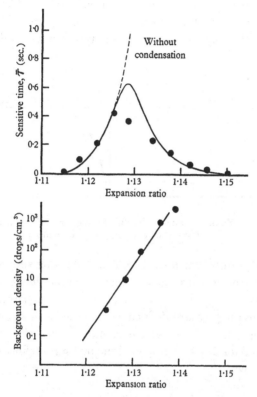

Fig. 10. Sensitive time and density of background condensation in a volume-defined chamber as a function of expansion ratio (Hazen).

method has been used by Joliot; Hazen, on the other hand, caused a pure β-ray source with a collimating shield to move in the chamber under intermittent illumination, and observed the positions of the source from which visible tracks proceeded.

Persistence of supersaturation in a pressure-defined chamber

The conditions which reduce supersaturation in a pressure-defined chamber are very different from those we have considered above, and the chambers themselves do not fall into a single category. If expansion depends upon the connexion of two vessels of finite size (§3.4), the operative effects follow closely those in a volume-defined system; on the other hand, if expansion is into the free atmosphere, the local heating of gas by conduction from the chamber wall or from condensing drops does not lead to any corresponding compression of the remaining gas volume; the persistence of supersaturation is thus longer, but less uniform.

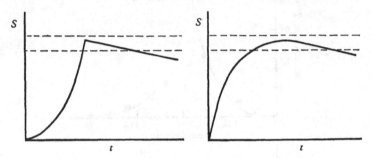

Fig. 11. Sensitive time as modified by the form of expansion
(left, volume-defined; right, pressure-defined).

A second feature which tends to make adequate supersaturation last longer in the pressure-defined than in volume-defined chambers is illustrated in fig. 11. The slow approach to maximum super-saturation in the pressure-defined chamber leads to an appreciable lengthening of sensitive time. An example of the relative sensitive times of comparable volume-defined and pressure-defined chambers is given below.

Chambers of prolonged sensitive time

The partial compensation of the reduction of supersaturation which is always to some extent present in a pressure-defined chamber can be deliberately fostered, and in this way chambers of long sensitive time—perhaps ten or twenty times the normal value—have been used by Bearden (1935), Frisch (1935), Meyer-Leibnitz (1939) and others. The apparatus described by Meyer-Leibnitz is shown schematically in fig. 12. Here the main expansion into the large receiving vessel is followed by a continued slow expansion into the smaller vessel at a rate which can be adjusted to maintain the level of supersaturation in the cloud chamber. The limit of the duration which can be reached by these methods probably lies in the essential instability of the heated gas at the floor of the chamber, although this may be minimized by making the floor a porous diaphragm through which the heated gas is withdrawn as expansion proceeds.

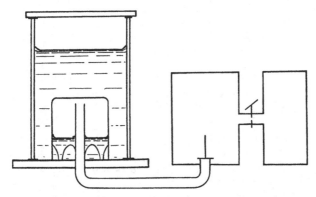

Fig. 12. Schematic diagram of a cloud chamber of prolonged sensitive time (Meyer-Leibnitz).

Typical sensitive times

In the expressions already given (equations (31), (32)) for the sensitive time, the size and geometry of the chamber is concerned only in the factor $(V/S)^2$. Unfortunately, it is often difficult to describe the area S accurately, since the porous diaphragm, if present, will act as a surface of the chamber wall but of uncertain area. The following examples (Table XI) are probably sufficient to allow estimates of sensitive time for other apparatus to be made.

The results given by Joliot show very clearly the reduction of sensitive time with decreasing pressure; the behaviour at the lowest pressures, where the sensitive time does not fall off as quickly as the total gas pressure in the chamber, is certainly a secondary effect due to the finite speed of expansion which is also shown up by the increased nominal expansion ratio which was found necessary at these pressures. The reduction of supersaturation due to evaporation is then so rapid that the variation of supersaturation even in a volume-defined chamber will more nearly correspond to fig. 11 (right) than to fig. 11 (left). Hazen's figures referring to a helium-

TABLE XI. *Typical examples of experimental sensitive times*

Workers	Chamber description					Comment
	Diameter (cm.)	Depth (cm.)	Gas	Pressure (atm.)	$\bar{\tau}$ (sec.)	
Williams	30	30	Air	1	0·4	
Williams and Evans	20	20	Argon	80	1·9–2·0	See footnote*
Joliot	Small	Small	Air	1·0	0·012	
				0·7	0·007	
				0·5	0·005	
				0·03	0·002	Pure water vapour
Hazen	30	10–22	Nitrogen	1	0·5	Volume-defined
			Helium	1	0·3	Pressure-defined
			Helium	1	0·7	
Meyer-Leibnitz	20	5	Propane	1	1–3	Artificially prolonged

* I am indebted to Dr G. R. Evans for this figure; the sensitive time of the chamber increases linearly with pressure up to about 10 atmospheres but only slowly at higher pressures. The natural rate of production of fresh ionization may be the essential factor limiting sensitive time in this apparatus.

filled chamber are the only ones available which seek to show the difference of persistence of supersaturation in a volume-defined and a pressure-defined chamber.*

(†A detailed study of simultaneous pressure and temperature variations at expansion has been carried out at Utrecht by Milatz and van Heerden (1947); van Heerden (1945) in a 'dry' cloud chamber (i.e. without condensant). These measurements exhibit clearly the characteristic behaviour both for volume-defined and for pressure-defined operation, and in addition the onset of heating by conduction in the central parts of the chamber. With a 'wet' chamber, confirmatory measurements of sensitive time were made using a rotating 'lighthouse' of collimated α-particles. For the rather small chamber used the pressure-defined sensitive time was about twice that obtained in volume-defined operation. Sensitive-time measurements were extended into the background-controlled condition (fig. 10), and, using different depths of volume-defined chamber, the transition of sensitive time from dependence upon chamber volume at low background to values independent of volume at high background was shown.)

3.8. Continuously sensitive chambers

All cloud chambers which have been brought into general use have depended upon the production of supersaturation of relatively short duration, and all the incidental techniques of chamber operation have referred to such apparatus. It is therefore difficult to assess with any confidence the possible advantages of a chamber in which supersaturation is maintained continuously at the level required for condensation on ions. New methods of control would no doubt be required; it is probable, for example, that in such an apparatus, the shower-selecting arrays of Geiger counters, which control the conventional chamber for recording the electron cascades in cosmic rays, would be replaced by photoelectric systems in which photographs are taken following sudden variations in the intensity of scattered light. As typical of the principle involved to obtain a continuously sensitive apparatus, we describe shortly that

* A comparison of the temperature rise in the centre of a cloud chamber (without condensation) for volume-defined and pressure-defined operation has recently been given by Endt (1948).

† Added November 1949.

due to Langsdorf (1939), which is probably the most successful continuously sensitive cloud-track apparatus which has yet been attempted.

The Langsdorf chamber

The Langsdorf chamber (fig. 13) consists of a shallow cylinder with axis vertical, parallel which a large temperature gradient is maintained, the enclosed gas is saturated at the upper (hot) surface

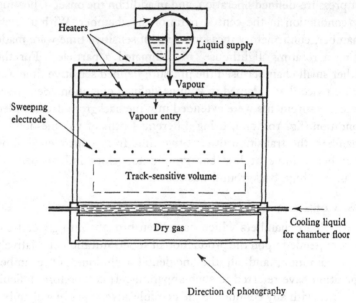

Fig. 13. Continuously sensitive chamber (Langsdorf).

and the lower (cool) one, and is accordingly supersaturated to an extent which may or may not be relieved by condensation in the intervening space. When the temperature difference between the upper and lower surfaces of the enclosed gas volume is made large enough, a region in which condensation takes place on ions is obtained, and this may be caused to occupy almost the whole lower half of the chamber. The large temperature gradient over the apparatus leads to considerable difficulties of mechanical construction. Other notable points are:

(1) The steady diffusing state must be stable under gravity. With the necessary large temperature ranges this is a severe

condition, demanding a light vapour and a relatively heavy gas. Methyl alcohol diffusing in carbon dioxide has proved successful. A typical mixture using a very heavy gas is of butyl alcohol diffusing in dichlorodifluoromethane; this is less satisfactory, for the slow rate of diffusion cuts down the permissible maximum condensation rate to an unsuitably low value.

(2) The vapour diffusing from the hot surface must itself be free from condensation nuclei. Langsdorf has shown that this result is achieved by drawing vapour from a free liquid surface heated by radiation from above.

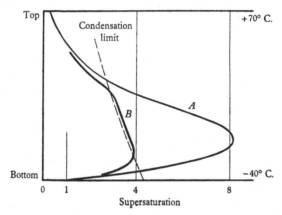

Fig. 14. Operating conditions in an apparatus of the Langsdorf type.

(3) A minimum of condensation other than on tracks is allowable in the sensitive volume, for the replenishment of vapour is limited by diffusion. Not only must the introduced vapour not carry condensation nuclei, but ions formed in the top half of the chamber (in which condensation on ions does not occur) must be prevented from reaching the sensitive lower half. This is the essential function of the electrostatic sweeping field, which is therefore maintained with respect to an open wire grid placed just above the upper limit of the sensitive volume.

The operation of the diffusion chamber is shown diagrammatically in fig. 14. The distribution of vapour through the chamber is determined by the one-dimensional diffusion equation, while the saturation pressure is determined by the corresponding temperature gradient; curve A shows the form of supersaturation distribution which then develops. The critical ion condensation curve is shown dotted, and a very similar curve would represent the onset of slight background condensation. When condensation occurs, the vapour pressure at all lower points in the chamber is reduced, and in equilibrium operation the actual supersaturation curve is probably of the form B, which will nowhere show much greater supersaturation than the dotted curve.

Although continuous sensitivity can be obtained in an apparatus of the Langsdorf type, the conditions which make the normal cloud chamber an instrument of precise measurement are not well met, and probably for this reason no effective application of the apparatus has yet taken place.

CHAPTER IV

OPERATION AND PHOTOGRAPHY

(1) BASIC PRINCIPLES OF OPERATION

4.1. Construction and materials

The subjects discussed in the following sections are of an elementary and empirical nature, and are intended to give the minimum general background necessary for routine chamber operation.

Constructional elements of normal chambers include windows, usually of glass or 'Perspex', metal parts, rubber for piston mounting and similar purposes, and sealing materials. These will be chosen to minimize the risk of contamination in the chamber leading to an unacceptable onset of background contamination. The information on this subject is almost wholly empirical. Glass and 'Perspex' are clean materials, and can be used freely, but special properties of 'Perspex' must be noted. This material absorbs some organic vapours including alcohol, and when this occurs its surface layers are softened. In these conditions the surface probably reproduces the favourable action of the gelatine layers used in the classical cloud-chamber researches of C. T. R. Wilson, but without any loss of optical properties. The softened surface, however, is most delicate, and on opening the chamber should not be touched until the absorbed vapour has evaporated completely. Metals such as silver, gold, platinum, chromium and nickel are clean, but the base metals are sources of definite contamination, and although this may be minimized if the metal is thoroughly oxidized, there is every reason for having all internal parts of the chamber plated. The qualities of aluminium are obscure; some workers find it satisfactory inside the chamber, while others claim it to be the seat of intense contamination. Rubber and neoprene are clean materials when at rest. In rapid motion and in particular under rapid changes of tension, such as occur at the support of the chamber piston, behaviour of great complexity is observed, leading, probably as a result of separation of charge between the surface of the material and the surrounding gas, to heavy background condensation. This behaviour can largely be prevented by designing the rubber parts as far as possible to move in constant tension. Sealing materials, such as vaseline and gold size, are clean, as are most enamels when thoroughly dried.

The chamber should be scrupulously clean when assembled, and in particular the windows should be completely free from grease. If this is not the case the unclean places become much more conspicuous in time and may make it necessary to take down the chamber. This deterioration of cleanness of windows is most marked if there is any tendency for free liquid to migrate to the glass surface on account of incorrect temperature gradients in the chamber.

4.2. Cleanness against background condensation

It is essential to be able to judge visually the condition of a cloud chamber. Not only must the chamber be satisfactory when first put up, but since it must be expected that the condition of the chamber will deteriorate with time, either by loss of vapour by diffusion out of the chamber or by slight chemical action leading to traces of contamination, the gradual deterioration for these reasons in use must not be overlooked. This may easily happen unless the chamber is regularly tested thoroughly in the way outlined below. The tests will be visual rather than by inspection of photographs, for not only is the use of photographs much too slow, but in photographic methods the details of chamber performance may be confused with faults of photography and of timing which can be diagnosed with certainty only if the condition of the chamber is first known. It is thus most important to equip apparatus with a source of continuous illumination, for visual inspection of condensation, which should be designed to come into operation with a minimum of alteration from the regular running conditions so that frequent inspection is encouraged. An intense source with good condensing system should be used, because for visual inspection it is essential that individual drops can easily be seen; even with good illumination this is preferably achieved by viewing at a small scattering angle, of say 30–45°.

The cleanness of a chamber is estimated by reference to (i) the development of background condensation over a range of expansion ratios in the neighbourhood of the ion limits, and (ii) to the way in which the residual nuclei effective at low-expansion ratios are removed in 'cleaning expansions'.

The cleaning process

Failure to clean satisfactorily under subsidiary cleaning expansions indicates either a leak of atmospheric air with its dust content into the chamber, or, more usually, the development of severe chemical contamination. The cleansing process in a good chamber is most characteristic: after heavy ion condensation the first one or two slow expansions appear to do little towards removing the re-evaporation nuclei formed; but after a certain point, cleaning proceeds rapidly, and a clean chamber, with less than 1 drop condensing in upwards of 10 cm.^{3} of gas, results. The successful later stages of cleaning take place when few enough nuclei remain for each to grow to a large drop which is likely to fall to the bottom of the chamber before it once more evaporates. This stage is particularly well marked in hydrogen, being assisted by a high diffusion coefficient and by low viscosity. The cleaning process thus allows it to be established that there is no continuous source of nuclei effective at low-expansion ratios in the chamber.

Condensation near the ion limit. This offers the real test of cleanness in the chamber, and it is necessary to establish a reliable subjective standard. A general test will consist of making expansions at gradually increasing expansion ratios from a level at which background condensation is negligible up to that at which it is certainly too thick. This is done in the presence of a weak γ-ray source which provides a few electron tracks, and the relationship of the development of background to the ion limits is in this way established. Although it is frequently not done, it is of the greatest value to carry out this series of tests in terms of a known calibration of expansion ratio. This procedure is of importance in two ways: it gives a measure of the range of expansion ratio for the transition from the first onset of condensation on ions to the point at which background becomes unacceptable, thus giving to some extent a non-subjective criterion of quality; secondly, it allows day-by-day changes of conditions to be followed and the development of contamination and of complete or selective drying out to be recognized. Under normal conditions, the range of expansion ratio from the onset of condensation on ions to the development of unacceptable heavy background condensation should be of the order 2 % (water vapour in oxygen) or 1 % (25/75 water-alcohol in argon).

It is most instructive to bring a clean chamber through a series of increasing expansions. The first indications of condensation on ions should come while background is still negligible. As larger expansions are made, the appearance of tracks becomes strikingly crisp and bright as the background drop density rises to about 1–10 drop/cm.[3] It is not easy to account for the improvement of appearance of tracks under these conditions, but it is undoubtedly the basis of the best visual judgement of chamber conditions. The important factor is probably that until almost 100 % condensation is taking place on ions of both signs, there will always be some ions which are fixed with appreciable delay, and that as long as this is so the secondary clusters, which are visually the conspicuous features of cloud tracks, will not appear as sharp points of light. In addition, at the early stages, there will only be half the final number of droplets present; the extra excess vapour pressure available at the higher expansions is negligible as an agent leading to brighter drops.

All stages of contamination will be encountered from the clean condition which has just been described up to a condition in which recognizable tracks are never obtained. The purpose of the last paragraphs is to stress the importance of being able to recognize good working conditions and to detect the intermediate levels of contamination at which tracks of a sort are formed which, however, at the acceptable limit of background are still in a condition corresponding to the early stages of the test sequence in a clean chamber.

Contamination: Remedies. When contamination is encountered it is often necessary to take the cloud chamber down and to work empirically until the source of contamination is identified and removed. With the constructional materials briefly referred to above, the most likely source of contamination is the solvent present in sealing compounds and paints, which are themselves clean when perfectly dry. It very often happens, however, that a chamber, initially clean, develops contamination in course of time; this may be ascribed to slight chemical action and can usually be removed without taking down the chamber. The chamber is swept with dry gas until the initial liquid is completely removed; fresh liquid is added and the chamber will as a rule be found to be clean.

4.3. Stability of temperature

There are serious difficulties in operating a chamber in surroundings where there are large fluctuations of temperature. These difficulties are of two kinds: the free liquid in the chamber tends to distil to the coolest part (if conditions are not controlled, sooner or later, this will be one of the windows), and further, temperature variations lead to convection currents in the chamber (§ 6.6 below) which prevent geometrical measurements of high precision. The chamber should therefore be operated in a room or enclosure of steady temperature; in addition, and particularly in vertical chambers, *slight* cooling should be applied at the lowest point in the chamber, and the free liquid should be introduced here. This local cooling is useful in two ways: it prevents the migration of liquid to unwanted places and it stabilizes the gas of the chamber, through which there is a negative temperature gradient in the upward direction. It is because of this stabilizing action that cooling must be used in moderation; for while stability at the time of expansion is important, the recovery of the chamber after the cleaning process may be greatly delayed if stability of gas is established too soon. After the last cleaning expansion, the chamber gas is compressed, and warmed. Before it is in a condition for further use, it must be saturated and must also come into a condition of stability. In the early stages, the restoration of saturation is assisted by strong convection, but later if stability is reached before saturation is virtually complete throughout the gas, the remaining transfer of vapour must be by simple diffusion. This premature stability is likely to arise if the cooling at the bottom of the chamber is excessive.

4.4. Automatic mechanisms

It is now usual for the greater part of the cycle of chamber manipulation to be carried out by an automatic mechanism. While the apparatus developed for this purpose is of great variety of design, a few salient features should be stressed.

As regards the chamber itself—ignoring, that is, the operation of cameras and other associated equipment—the zero of time in the control-mechanism cycle depends on the nature of the initiating signal; in work in the nuclear field the signal is in the control of the operator and may be synchronized with the admission at the instant of full supersaturation of ionizing particles; in counter-controlled work the signal comes at the instant of passage of uncontrolled particles. The important features of the mechanism are the same in each case:

(a) the timing of the illuminating flash should be continuously variable from (say) 0·01 to 0·5 sec., measured as the case may be from the admission of the particle beam or from the operation of the counter control (§ 5.2 below);

(b) the number of cleaning expansions should be readily variable over a considerable range, say from one to ten;

(c) a delay mechanism must prevent operation of the chamber after the cleaning cycle is completed to allow for the restoration of saturation and for gas stability to be attained. The necessary delay must be found by experiment, and a controlled delay, a thermal switch, for example, is suitable.

(2) OPTICS AND PHOTOGRAPHY OF CLOUD DROPLETS

4.5. Nature of object

A cloud track offers two distinct types of object for photography. Single water drops, of diameter of the order 10^{-3} cm., give images of size and shape determined entirely by instrumental factors, by the diffraction disk of the lens aperture, by the lens aberrations and by the resolution of the photographic emulsion. The small close clusters of drops, on the other hand, formed by slow secondary electrons, yield images which reproduce to some extent the shape and scattering power of the cluster. Since these extended images are essentially the result of superposition of single-drop images, it follows that they are more easily photographed, and under poor conditions they alone appear in the photograph. The tracks of α-particles, slow protons and of heavier nuclei are extreme examples of drop clusters.

The level of technique which is required depends on whether individual drops are or are not to be recorded. It has become apparent in recent years that for critical work there are strong reasons for working consistently to the standard of single-drop recording. Single drops define more clearly than do clusters the actual trajectory of the primary particle, for no point in a close cluster, even if there were no extension due purely to diffusion, can properly be identified as more likely to be the point of origin, and so to lie in the primary trajectory, than another. Further, if ionization, either total or primary, is to be measured, and also more generally in exploratory photographs, when the interpretation may depend on approximate ion counts, on curvatures measured over very short lengths of track, or on the presence or absence of recoil spurs of a very few drops, the certainty that every drop is to be seen is of the first importance.

We therefore discuss the problem of track photography in general from the standpoint of recording individual drops.

4.6. Light scattered from droplets

At the time of photography, the droplets of background condensation and of thinly ionized tracks are of radius of the order 20 wave-lengths, while those in dense tracks, in which there is competition for the vapour supply immediately available, are considerably smaller. Drops which must be photographed as isolated objects will always be of the larger size.

The angular distribution of scattering by these drops has not been expressed in a form suitable for calculation; the formal expressions given by Mie (1908) and Rayleigh (1910) become quite unmanageable for drop radii greater than about 2 wave-lengths, while the results of geometrical optics are not applicable until considerably larger drop sizes are reached. The general form of distribution is not in doubt, however, and a valuable experimental study of the important range of angles was made by Webb (1935).

There is an intense maximum of forward scattering from which the intensity falls continuously to a minimum at scattering angles rather greater than 90°; beyond this minimum the scattered intensity increases again to a feeble maximum in the backward direction.

Webb's measurements, which refer to water drops and those condensing from the normal water-alcohol mixture, and to drops of roughly the diameter quoted above, are given in Table XII, and are plotted in fig. 15. The very great increase of scattered intensity in the forward direction, certainly of 20 or 30 times between 90° and 30° scattering, is well brought out, but it is likely that great stress should not be placed on fine details of the table. The drops of each condensant are of similar size and, in fact, give similar total scattering; their refractive indices cannot be very different. It is therefore difficult to understand how alcohol-water drops at scattering angles near to 90° should be appreciably more effective scatterers than drops of pure water. When particular care is taken to photograph at the earliest possible instant, in order to minimize geometrical distortions, the drops at the instant of photography will probably be distinctly smaller and the variation of scattered intensity with angle will be rather less marked. A large amount of cloud-chamber photography is carried out at a scattering angle of about 90°, which we have seen to be a direction of decidedly inefficient scattering. While this choice is sometimes dictated by geometrical features, for example, curvature in a magnetic field is necessarily photographed along the direction of the field, its wide adoption is based on two general features, the plate-like form of the region of good focus (which is conveniently taken to occupy a large part of a shallow cylindrical chamber), and the fact that only under right-angle illumination is the physical background of photography and the window through which the tracks are photographed easily left free from light. On the other hand, it has been shown to be possible to work at very much smaller angles without restriction of the space illuminated, provided that the windows through which light enters the cloud chamber may be in the field of view of the camera (see, for example, Wilson and Wilson (1935), in which an angle of scattering of about 20° was used). However, this procedure requires that these windows shall be kept exceptionally clean, and this presents serious difficulty.

For many practical purposes, however, small deviations from 90° scattering are feasible in conventional chambers, and it will be found that deviations as small as 15 or 20° lead to marked increases of scattered intensity of which advantage can often be taken.

TABLE XII. *Scattered intensity from water drops*
(diameter ∼ 2·10⁻³ cm.)

Angle of scattering (°)	Relative scattered intensity	
	Water	Alcohol-water
20	100	63
40	12	13
60	2·5	4·7
80	0·8	2·0
100	—	1·0

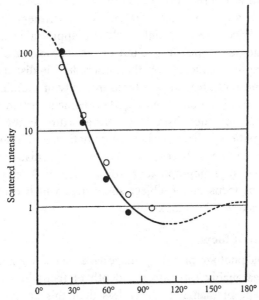

Fig. 15. Scattering of light by cloud-chamber droplets. The full line covers the region investigated by Webb (1935) (● water, ○ alcohol-water), the broken curve is conjectural.

4.7. The photographic image of a single drop

Inspection of a photograph on which single drop images can be distinguished shows immediately that all of the images are of approximately the same size. In fast emulsions the diameter of the drop image is between 2×10^{-3} and 3×10^{-3} cm., and the image is thus considerably larger than the purely optical image in the plane of focus. The size of the photographic image is therefore set entirely by the properties of the emulsion, by grain size and turbidity. It follows that the conditions for recording a drop image are that:

(*a*) the camera lens must collect enough scattered light from the drop to provide the necessary threshold intensity spread over an image disk of diameter, say, 2×10^{-3} cm., and

(*b*) it must focus this light into an optical image disk which is of appreciably smaller diameter.

The degree of blackening required in such small, disk-like images before they can readily be identified is, of course, very much greater than would be required to differentiate extended areas, and is essentially a quantity to be determined experimentally. Only if there is a large excess of light will any appreciable number of larger images, corresponding to out-of-focus drops, be observed.

Since the light collected by the camera lens is distributed over a disk of constant area, the scattered intensity of illumination and the properties of the emulsion together define a certain minimum solid angle, subtended by the camera aperture at the drop, for image formation. Hence, when illumination and photographic equipment have been chosen, there is a maximum distance at which it is possible to photograph single-drop images. At this distance, the depth of focus in the object space over which drops can be photographed will be a maximum.

4.8. Depth of focus

If the final photographic drop image has a diameter of 2×10^{-3} cm., we may reasonably estimate the depth of focus in terms of an optical image of diameter rather less than this of, say, 10^{-3} cm. This is too small an image circle to be treated by purely geometrical optics, and we will use the approximation that the lens is free from spherical aberration, and that therefore the image circle substantially retains the diffraction diameter of the true focus up to a distance of about twice the Rayleigh limit from the focus, that is, until the path difference between central and extreme paths corresponds to a variation of phase of π, and that, further, the image disk outside this limit increases approximately geometrically. It must be stressed that this is an ideal situation, and that wide-aperture lenses, particularly when used at low magnifications for which they have not been specifically corrected, are likely to show considerable departures from it.

If the lens aperture is $2a$ and the image distance u, the disk-like diameter of the central diffraction circle may be taken as about

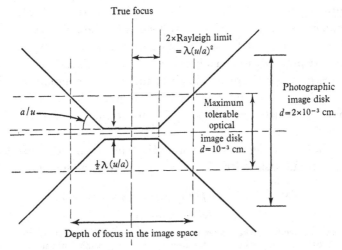

Fig. 16. Diagram illustrating the relation between the Rayleigh region of focus and the photographic image disk of a single drop. Lens $f/2$, free from spherical aberration, the transverse scale of the diagram is 4 times the axial scale; $\lambda = 5 \times 10^{-5}$ cm.

$u\lambda/2a$, while the Rayleigh limit of focus, corresponding to a path difference of $\frac{1}{4}\lambda$ between central and extreme rays is $\delta u = u^2\lambda/2a^2$. The intensity of the central diffraction disk varies little up to the Rayleigh limit, while its diameter is practically unchanged up to the double limit, $\delta u = u^2\lambda/a^2$, where the disk intensity is 0·45 of maximum. We make the assumption that the integrated intensity over the image disk does not vary over the further region, outside the double limit, which is relevant.

The relations between photographic image disk, optical disk, disk of perfect focus and the Rayleigh limit, are illustrated for a lens of aperture $f/2$, free from spherical aberration, in fig. 16. It will be observed that about half the total depth of focus lies in the region of the minimum image disk, a proportion which increases for lenses of smaller aperture. This may well be a fact of considerable importance, for if photographic materials of adequate speed and of much greater resolving power were available, very much smaller drop images might be expected with little reduction of depth of focus, and detailed drop counts might then be possible on normal counter-controlled photographs. Lenses showing spherical aberration under the conditions of use will, in general, have a smaller depth of focus, but the extent of loss is a matter of detail. Table XIII summarizes the values of depth of focus which are obtained by the procedure outlined for typical apertures and focal lengths, and these are probably reasonably correct for materials now in use. The figures in brackets are the depth of focus corresponding to twice the Rayleigh limit, and thus the depth of perfect focus with photographic material of very high resolution, and with a lens free from aberration.

Although the figures in Table XIII refer to an ideal lens, they offer a useful basis for the discussion of practical matters concerning

TABLE XIII. *Depth of focus in the object space for drop images*

Figures outside brackets refer to normal photographic emulsions in current use; those within brackets are limiting values for possible emulsions of very high resolution (§ 4.8). ($\lambda = 5 \cdot 10^{-5}$ cm.).

f number ...		2	3·5	5·6	8
Minimum optical image diameter (cm.) ...		10^{-4}	2×10^{-4}	3×10^{-4}	4×10^{-4}
2 × Rayleigh limit (cm.)...		$0 \cdot 8 \times 10^{-3}$	$2 \cdot 5 \times 10^{-3}$	$6 \cdot 2 \times 10^{-3}$	$1 \cdot 3 \times 10^{-2}$
Focal length (cm.)	Object distance (cm.)	Depth of focus in the cloud chamber (cm.)			
3·7	30	0·3 (0·1)	0·6 (0·3)	1·1 (0·7)	1·8 (1·3)
	60	1·3 (0·4)	2·5 (1·2)	4·9 (3·1)	8·4 (6·1)
	100	3·6 (1·1)	7·4 (3·5)	14 (9·0)	25 (18)
5·0	50	0·4 (0·1)	0·8 (0·4)	1·6 (1·0)	2·9 (2·1)
	100	1·9 (0·6)	3·8 (1·8)	7·3 (4·6)	13 (9·2)
	200	8·0 (2·5)	16 (7·7)	32 (20)	55 (40)
7·5	50	0·2 (0·1)	0·4 (0·2)	0·6 (0·4)	1·1 (0·8)
	100	0·8 (0·3)	1·7 (0·8)	3·2 (2·0)	5·5 (4·0)
	200	3·5 (1·1)	7·0 (3·3)	14 (8·6)	23 (17)

the choice and use of suitable lenses. We will assume throughout that it is necessary to photograph single-drop images.

(1) *Depth of focus with given lens.* With given illumination the maximum depth of focus for which individual drops are photographed is obtained at the maximum aperture of a lens. For example, a 3·7 cm. lens at $f/3·5$ has depth of focus 7·4 cm. at 100 cm. If the aperture is reduced to $f/5·6$, the object distance at which the lens aperture subtends the same solid angle at the drop to be photographed is reduced to 63 cm. and the depth of focus is only 5 cm. At $f/8$ the object distance becomes 44 cm. and the depth of focus 3 cm. Thus, with a given lens and illumination, magnification necessary for the separation of drop images, if these are to be counted, is purchased at the price of depth of focus. Further, as aperture is reduced and magnification at the limit of drop recording is increased, the area of chamber covered by the lens is reduced.

(2) *Lenses of differing focal length used at the same magnification.* It is frequently desirable to define the whole photographic system with regard to a particular negative size, most often the size permitted on normal 35 mm. ciné film. Then the object distance is roughly proportional to the focal length of the lens, and lenses will reach the limit of drop recording at equivalent apertures. There is a slight advantage, as regards depth of focus, with the short-focus lens. It is doubtful whether this is large enough, however, to outweigh the practical advantages of using a long-focus lens for some purposes, for example, when a cloud chamber contains a system of plates which are conveniently all observed approximately edge on. The variation just described is much slower than that in (1) above; it follows that (1) describes approximately the behaviour of depth of focus with magnification whether or not the same lens is used.

(3) *Light utilization under counter-control.* As is discussed in a later section, the conditions in the exploratory photographs which are of particular importance in cosmic-ray research to-day offer a variety of ways in which a given intensity of illumination can usefully be applied, and the conditions actually used must be a balance between competing claims. For example, suppose that photographs are being taken at a given level of illumination, and that the intensity of available light incident on the object drop is increased. The extra light may be applied (*a*) to give greater depth of focus at a given

magnification, thus allowing a larger volume of chamber to be surveyed in each photograph, (b) to increase the magnification, thus increasing the separation of drop images, and allowing more accurate conclusions about ionization density in tracks, (c) to permit an emulsion of high resolution (but slower) to be used to the same purpose, (d) to increase the diffusion time of tracks without decreasing the precision of geometrical measurements. It is interesting to notice that three alternative methods are suggested for increasing the accuracy of drop counts without loss of geometrical precision. The second of these, which involves an actual reduction in the diameter of drop-image disks has not been demonstrated in practice, and its potentialities may well be limited by the standard of performance of optical equipment readily available. The other two are more interesting, but information at the moment is lacking which might help to distinguish between them.

Very little general comment on the balance of light utilization is possible. There is no doubt, however, that in practice the factor most frequently ignored in this balance is the growth time of droplets up to the instant of photography. The objectionable features of long growth time may scarcely show up at all except in a prolonged series of geometrical measurements, and even then only against some poorly known standard of precision. Meanwhile, on account of the strong scattering of overgrown drops, the other features of the photographs are admirable, strong drop images being combined with a satisfactory depth of focus. It is important to remember that the growth time is likely to be at least as long as the expansion time from the passage of controlling particles until expansion is completed, and that the growth time is therefore at least as important in setting the limits of distortion by the usually dominant steady convection currents of the chamber (§ 6.6 below). In an example recently given by Adams and others (1948), the growth time is nearly 20 times the expansion time, but this example is abnormal in so far as it refers to an exceptionally fast chamber, while the growth time could probably have been reduced at the expense of other factors involved in the supply and use of the illuminating flash.

CHAPTER V

COUNTER-CONTROL

5.1. The need for counter-control

Control of the operation of a cloud chamber by an array of
Geiger counters concerns so large and important a section of cloud-
chamber work that it may properly be considered as one of the
basic techniques of chamber operation. It is particularly in cosmic-
ray investigations that, up to now, control has been needed, and it is
in this context that we consider the control problems.

The need for counter-control can be brought out by first
estimating the rate at which typical cosmic-ray material might be
obtained by random photography, that is, by operating a cloud
chamber as often as the recovery time between successive expansions will allow. The success of this policy will be found to depend
entirely upon what can be regarded as a successful photograph of
a track or group of tracks. For a few problems (examples are the
early demonstrations of meson decay) photographs of very moderate
quality, quite unsuited to precise measurement, are sufficient to
show the essential features of the event, and it is under these conditions that random photography shows to the best advantage, since
use may be made of the full period of chamber sensitivity, and
chambers of long sensitive time may be used.

These applications, which require only qualitative results from
the cloud-chamber photograph, are not common, however; indeed,
it is precisely in cosmic-ray investigations, for which control has
been applied, that the question of accuracy of measurement (which
is dealt with in some detail in the next chapter), has arisen. It will
be apparent (§6.5) that the time during which tracks can enter
a cloud chamber and be photographed with distortions not greater
than those of a track actuating counter-control is only of the order
of the expansion time, say between 0·01 and 0·02 sec. This collecting
time, within which the tracks of particles entering the chamber at
random are distorted in a similar degree to counter-controlled
tracks, can probably be lengthened, perhaps to 0·05 sec., before
appreciably increased distortion is to be found, but the use even of

this time as the collecting time of a cloud chamber as a precision instrument will show clearly the limitations of random operation.

The flux of ionizing cosmic-ray particles at sea-level is approximately 1 particle per horizontal square centimetre per minute, with a cosine-squared distribution about the vertical. Thus the particle flux through a chamber 30 cm. diameter × 5 cm. deep will be of the order 3 per second, while through a larger chamber it will be correspondingly greater. Then, since the sensitive time of a normal chamber of the dimensions quoted may well be $\frac{1}{5}$ sec., about every other photograph will contain some part of the track of a cosmic-ray particle, and a moderate increase of size or pressure will bring in a track for practically every photograph.

Additional restrictions, however, on the direction of the particles, to particular categories of particle and to the much shorter time for which trajectories may be measured accurately, alter these figures of efficiency by an order of magnitude. A correct criterion of the efficiency of a cloud chamber operated at random is the fraction of the total time of operation within which it accepts and records tracks conforming to certain requirements of geometry and quality for measurement. This fraction determines the effective time for which the apparatus has actually been available to receive a particular phenomenon under conditions leading to satisfactory measurement.

The chamber considered already, 30 cm. diameter and 5 cm. deep, will probably give good tracks, sufficiently undistorted, for most purposes, for $\frac{1}{20}$ sec., and will take perhaps 4 min. to reset after each expansion (for work of the highest precision the useful receptive time must be still shorter). The fraction of operating time for which the chamber is effectively receptive is then $\frac{1}{5000}$. Particles traversing the full depth of the chamber, say in a solid angle restricted to $\frac{1}{10}$, will occur at the rate of about 10 per minute, and hence will be photographed, if the apparatus is in continuous operation, once in 8 hr. Particular particles present to 1 % of the whole flux will be recorded only once per month of continuous operation, while such events as the penetrating showers described by Janossy and others, which occur at sea-level approximately once per 10^5 particles, will not be photographed at the required level of quality oftener than once in 80 years!

The operation of a cloud chamber at random is of definite value as the most effective way of exercising an unbiased selection of particles, but the rate at which precise data of scarce but recognizable events will be accumulated is extremely low. The control of a cloud chamber by counters, introduced by Blackett and Occhialini (1933), decisively extended the range of events which can be photographed at a useful level of quality.

5.2. Technical problems of counter-control

The basic technical problems of counter-control have been considered in detail by Blackett (1934); they depend on the time available after a particle or group of particles has operated a Geiger-counter array, during which the resulting ions may be fixed by condensation in the cloud chamber sufficiently accurately on the original line of formation.

The diffusion of the thin cylindrical column of ions left along the trajectory of a fast particle traversing the gas of the chamber may be given in the form

$$n(r) = \frac{N_0}{4\pi D\tau} e^{-r^2/4D\tau}, \tag{35}$$

when N_0 is the total number of ions of both signs per centimetre of track, D is the diffusion coefficient of the ions, and $2\pi r\, n(r)\, dr$ is the number of ions in the cylindrical shell, $r, r+dr$ per centimetre length at time τ after formation; recombination of ions is ignored.

In a track photograph we project on a plane, and when, as very frequently happens, this plane is approximately parallel to the original trajectory, the number of drop images per centimetre at distances between x and $x+\delta x$ from the centre in the projection is

$$\frac{N_0}{4\pi D\tau} \delta x \int_{-\infty}^{\infty} \exp\left[-\frac{x^2+y^2}{4D\tau} \right] dy,$$

while the density of drop images, $\rho(x)$, at a distance x from the centre of the image, is

$$\rho(x) = \frac{N_0}{\sqrt{(4\pi D\tau)}} e^{-x^2/4D\tau} = \rho_0 e^{-x^2/4D\tau},$$

where
$$\rho_0 = \frac{N_0}{\sqrt{(4\pi D\tau)}}.$$

The 90 % breadth of the image, i.e. the breadth X which contains 90 % of the total number of drop images, is given by

$$X = 4 \cdot 68 \sqrt{(D\tau)}. \qquad (36)$$

In air at N.T.P., $\qquad D = 0 \cdot 034 \, \text{cm.}^2 \, \text{sec.}^{-1}$,

and so $\qquad\qquad X = 0 \cdot 86 \sqrt{\tau} \, \text{cm.}$;

thus, if the tracks must not be more than 1 mm. broad, τ must not be greater than about $\frac{1}{70}$ sec.

The significance of this particular figure is not always appreciated; if the direction and curvature of tracks are not to be measured, a much greater time of diffusion is tolerable; on the other hand, a reduction of time will certainly reduce the governing errors of measurement to some degree. The interval $\frac{1}{70}$ sec. is to be taken as giving, to order of magnitude, the most rapid expansion which can readily be obtained using conventional mechanisms on chambers of normal size, that is, of a few litres capacity. It is determined by such quantities as the order of pressure difference which is available to drive the piston or diaphragm and the velocity of efflux of gas which may be attained from the back of the chamber. (The most rapid chamber operation which has yet been used is described by Anderson and his school (Adams *et al.* 1948) who expand a rather small chamber (17 cm. diameter) in 0·004 sec.) To complete expansion in much shorter time will require a considerable increase of complexity of expansion mechanism, while detailed problems concerning the arresting of the piston at the end of its motion and the electrical properties of rubber under rapidly varying tension, which even under present conditions can be severe, might well become uncontrollable. It is therefore important to see as clearly as possible the order of advantage which may be expected from such rapid expansion.

We shall see (§ 6.5) that the precision of geometrical measurement is time-sensitive when it is limited by chamber distortions, of which the steady convection drift of gas in the chamber is usually the most important. This drift is effective from the instant of passage of the fast particle, until the instant of photography; we may write this interval $\tau_e + \tau_p$, where τ_e is the time of expansion measured from the passage of the controlling particle, and τ_p is the time of growth of the cloud droplets up to the instant of photography. The interval τ_p is in some degree under control; photography may be advanced to the earliest possible moment and, by making more light available, the time of growth necessary may be reduced, since the scattered intensity from a growing droplet increases approximately linearly with time (§ 1.8). In practice, however, it is rarely possible to make any increase of light intensity available for this purpose. Even with modern illuminants and photographic emulsions there is no great excess of light in hand, and what there is will as a rule be used to take advantage of an increase of depth of focus by stopping down camera lenses. The growth time, τ_p, is thus for most purposes a constant which usually lies between 0·1 and 0·02 sec., and it is clear that no great advantage will arise from a reduction of τ_e when this is already appreciably smaller than τ_p.

We shall also see (§ 6.12) that the precision with which a trajectory is described by a diffuse band of drops is proportional to $\tau_e^{-\frac{1}{2}}$, and that when chamber conditions are good, the statistical uncertainty arising in this way is only a little smaller than that arising from chamber distortions. If τ_p were in fact much smaller than τ_e, any large reduction of τ_e would leave the statistical description of a trajectory as the governing uncertainty of measurement, and a reduction of time would only increase precision as $\tau_e^{-\frac{1}{2}}$. However, for existing values of τ_p, the effect cannot be expected ever to become of main importance.

A much more rapid expansion is thus of little advantage until the necessary time of drop growth can be greatly reduced. In fact, to increase the precision of measurement in counter-controlled tracks, the latter is probably as a rule the more fruitful point of attack.

5.3. The resolving time of a cloud chamber

The counter system controlling a cloud chamber may require a group of particles to pass through the chamber simultaneously to within a time of the order 10^{-6} sec., the resolving time of the counter-array and its associated equipment. These particles will be supposed to be strictly contemporary, although in counter-practice methods of estimating the possibility of random association of particles in terms of the resolving time of individual counters and their counting rate are well known. From the point of view of the chamber all of these counter-selected particles are exactly associated in time. It is quite possible, however, for random particles also to pass through the chamber during its effective collecting time, and it is necessary, when the causal association of particular particles is of importance, to be able to estimate the possibility that such random particles can be indistinguishable from the counter-defined group; to know, in fact, the resolving time of the cloud chamber.

Tracks entering the chamber after supersaturation is established are practically indistinguishable, and the resolving time for sharp tracks is thus either the sensitive time or a time determined by the instant of photography, whichever is the shorter. The time elapsing between the entry of a pre-expansion track and the establishment of supersaturation, on the other hand, determines the diffusion breadth of the track, and the breadth of non-contemporary tracks may be distinguishable.

We suppose that the breadths of two tracks will be distinguished if they are in fact different by a factor f. Under good conditions f may be as large as $0 \cdot 7$, but if the general technical level of the photograph is poor, or if there is any question of unevenness of illumination between the tracks compared, f will probably be distinctly smaller. Now, if τ_e is the time between the passage of the particle which operates the chamber and the fixing of its ions by condensation, the time during which particles can pass through the chamber yielding tracks indistinguishable in time extends from $f^2 \tau_e$ to τ_e/f^2; and t_0, the resolving time of the chamber, is thus

$$t_0 = \left(\frac{1}{f^2} - f^2 \right) \tau_e. \tag{37}$$

If $f \sim 0 \cdot 7$, $t_0 \sim 1 \cdot 5 \tau_e$, and more generally, t_0 is likely to range from τ_e to 4 or 5 τ_e according to the quality of work.

This elementary analysis is probably adequate for normal conditions of counter-control, when τ_e is of the order $\frac{1}{30} - \frac{1}{100}$ sec., although at much shorter times the finite breadth of 'sharp' tracks would become of importance. It is clear that the requirements of counter-control for precision measurement are precisely those which lead to the optimum resolving time, and that in random photographs the collecting time over which the resolving time is comparable with that under counter-control is only of the order of the time of operation of the chamber (see, for example, § 5.6 below).

5.4. Application of control: operating rate

The counting rate of an array of Geiger counters in the cosmic-ray beam depends broadly on two features: there is a geometrical factor, in so far as the counting rate depends upon an area, and perhaps upon a solid angle, of collection, and there is a selective factor in so far as the array only responds to certain of the particles, or sets of particles, falling upon it. The selective feature usually operates by exclusion; to a simple array which responds to many categories of events further elements are added which exclude the unwanted categories and leave only those required. Only rarely is a counter array specific to a single type of event. Selection by exclusion would ideally be abrupt, all desired events being passed and all others excluded, but in practice the selection will only be partial, and if the rigour of the conditions imposed is increased until hardly any unwanted events pass, the efficiency of passage of the required events may suffer severely. Used in isolation, a counter array designed to study a particular phenomenon must as a rule give a rigorous selection so that the properties examined are certainly features of the phenomenon under investigation, and a consequent reduction of efficiency of collection of wanted events may have to be accepted. (For example, see Janossy and Broadbent (1948), where the efficiency of a 'penetrating shower set' is largely reduced as a result of complications designed to exclude other events.)

When the array is used to control a cloud chamber, the situation is somewhat changed. There are still occasions when the full rigour of selection will be sustained, and these may be grouped as experiments in which the information of the cloud-chamber photographs is to be taken on a more or less equal footing with information deduced from the operation of a particular counter-array. More often, however, the cloud-chamber photograph will, when obtained, be taken on its own merits, and the interpretation will not stress the counter-operation by which it was obtained. Under these circumstances, an increased rigour of selection will, other things equal, merely reduce the total number of photographs obtained of the required event. We therefore indicate the considerations which must govern the counting rate appropriate to a counter-system controlling a cloud chamber.

Characteristic time

The recovery time of the cloud chamber is the time during which cleaning cycles are made after an operation and the period during which the chamber must remain inactive after resetting for the induced convection of compression to die away and for full saturation to be established throughout its volume; it is a characteristic time against which the operating rate set by a control system must be considered. It varies from 1 or 2 min. in a small chamber to 10 min. or even more in a large chamber, and to even longer in high-pressure chambers.

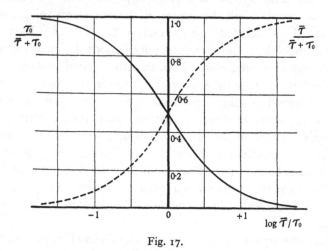

Fig. 17.

If we write τ_0 for the recovery time of the chamber, and $\bar{\tau}$ for the mean waiting time set by the counting rate, the rate of photography is $(\tau_0 + \bar{\tau})^{-1}$, while the fraction of all time for which the apparatus is receptive to the events sought is $\bar{\tau}(\tau_0 + \bar{\tau})^{-1}$. It is clear that the rate of photography cannot be greatly increased when $\bar{\tau}$ is appreciably smaller than τ_0, while the total receptive time cannot become much greater when $\bar{\tau}$ is appreciably larger than τ_0 (fig. 17). A high counting rate and a high fraction of receptive time are thus incompatible, and there is a basis for distinguishing between *common events*, for which a high collecting rate is of more importance than a high fraction of receptive time, and *rare events*, for which the maximum receptive time alone will provide for efficient collection.

Common events. Here $\bar{\tau}$ can easily be made much smaller than τ_0. For example, a counter telescope of two counters 40 × 4 cm. in a vertical plane 40 cm. apart will count mesons at a rate of about 8 per minute, while a chamber occupying the intervening space will have a recovery time, τ_0, of at least 3 min. Thus a rate of photographing of $60/(3 + \frac{1}{8}) = 19$ per hour is obtained. This is reached, however, by using the rather large depth of focus, 4 cm., and by having counters so long, compared with their separation, that tracks at inclinations up to 45° are included, which are probably more distorted than almost vertical ones. If $\bar{\tau}$ is increased to 1 min., the rate of photographing is still 15 per hour, but the accepted particle beam can be much more strictly defined; it may now be set, for example, by counters 2 × 30 cm., and the general standard of material photographed will be improved accordingly. (Since the more inclined tracks are now excluded and, with smaller depth of focus to cover, a larger lens aperture may be used and earlier photography of drops be achieved, measurements of momentum may be made more accurately.) The first, more rapid rate of photographing will certainly yield many fewer than 15 photographs per hour conforming to this stricter selection. This example shows the way in which for common events it is advantageous to manipulate geometrical factors until $\bar{\tau}$ is not very much smaller than τ_0.

Rare events. A clear-cut example of the treatment of rare events cannot be given. Broadly, two solutions exist, according as the rare event is selected with certainty or is selected with progressive loss of efficiency as the rigour of selection is increased. In the second situation, it will certainly be profitable to relax the rigour of selection until $\bar{\tau}$ is not too great, say $3\tau_0 < \bar{\tau} < 10\tau_0$. In the first situation, however, any relaxation will be unprofitable and no increase of the rate of photography of the desired events can be achieved. If $\bar{\tau}$ is still large (say $\bar{\tau} > 10\tau_0$), however, another quite independent experiment can be superposed upon the same apparatus, and provided that the total counting rate is still not too small ($\bar{\tau} > 3\tau_0$), each will be carried out efficiently; the apparatus will be receptive to either category of event for a large fraction of the total time. When experiments are superposed in this way, the photographs, if they must be distinguished, are marked by means of indicator lamps.

5.5. The counter-controlled chamber as a sampling device

Recent work by Cocconi and his collaborators (1946), Auger and Daudin (1945), Janossy and Broadbent (1948), and others has directed attention to the statistical problems which arise when counter-arrays are subjected to showers of particles distributed at random and of a variable density which may be assumed to be unchanged over areas large compared with that covered by the extension of the counter-array. The probability of operation of a given counter-selection can then be stated in terms of the mean particle density in the shower and the projected area presented by the various counters to the direction of the shower. By varying the areas of individual counters (most simply by using more or fewer unit counters in parallel), the density spectrum of the particle showers may be determined. Provided that the effective collecting area of a cloud chamber can be stated in the same way as is that of a counter, a chamber operated in coincidence with such a counter-array is a sampling device which gives an estimate of the density of individual showers which is limited only by the statistical fluctuations of particles within the shower and of the number of shower particles which pass through the area defined by the chamber.

A cloud chamber has been used in this way by Miss Chaudhuri (1948) in an investigation of the relationships between the production of penetrating effects under a thick absorbing screen and the local density of cosmic-ray showers from the atmosphere.

5.6. Counter-control in work at high altitudes

Two general technical features of great interest arise in the application of counter-control at high altitudes and in particular for cloud chambers operated in aircraft, and stress the interrelation of the control problem with other aspects of chamber operation. We will not comment on the detailed features of equipment, since, while the installations will be of a familiar kind, the details will be largely dictated by the types of aircraft to be used and the conditions under which these are made available.

The general problems are:

(1) the modification of chamber design arising from the fact that the particle flux is much more nearly isotropic at high altitudes than at sea-level, and

(2) since weight limitation may be overriding, and the choice may lie between a continuous but low magnetic field, and a field flashed for short periods, ought counter-controlled or random operation to be preferred?

Design related to particle flux

An empty, circular chamber presents no problems; if the object is to study the distribution of ionization density or of momentum in a particle sample, orientation is largely irrelevant, although (§6.11) momentum is still likely to be most accurately measured for vertical tracks, while a uniform electrostatic field perpendicular to the line of sight, necessary to separate the positive- and negative-ion columns in order to establish the existence of 100 % complete condensation, must necessarily fail to separate a track parallel to the field.

It is likely, however, that high-altitude experiments will mainly be directed to crucial problems such as that of meson formation, and it may be assumed that these will involve the use of chambers containing rather thick absorbing layers, probably more than one such layer.

It then becomes necessary to concentrate on a certain cone only of the cosmic-ray beam, for only a limited cone of angles of trajectory will traverse the plates in a satisfactory way. Unless tracks in this category appear in a reasonable proportion of random photographs, counter-control will be necessary. Further, the work will be concerned with the *association* of particles; hence the main value of any photograph will lie in tracks of an epoch when differences of age are most obvious—that is, in tracks entering the chamber just before expansion—for the difference which can be detected must be very much less than the mean time interval between the entry of independent particles; it is essential that the chance of an association of several particles at random within the resolving time (§5.3) of the chamber must be small. It follows that even in the absence of magnetic field, if we concern ourselves with informative tracks rather than with all recorded tracks, counter-control is likely to be fruitful.

Duration of field excitation

We first observe that the distinction between a continuously excited field used with counter-control, and a short duration field used with random operation, has very different force at high altitudes from that at sea-level.

The choice is only at issue if random photography does in fact lead to an appreciable proportion of tracks in a suitable geometry and, as we have seen, in a rather short time interval, which we may estimate (§ 5.1), as $\frac{1}{20}$ sec. (In fact, this figure is probably an over-estimate, for, since the chamber distortions are at least linear functions of track age, to double the time interval is equivalent to reducing the magnetic field to one-half or less.) This being so, if random photography is conceivably possible, the time interval of waiting ($\bar{\tau}$ in § 5.4) before operation of a counter-control cannot well be greater than about 1 sec. The alternative is, then, to pulse the magnet coils for a time at maximum field of the order $\frac{1}{20}$ sec. for random operation, or to do so for about 1 sec. for controlled expansion, for the field clearly need not be maintained in either case during the resetting and cleaning operation.

To summarize: if high-altitude chamber work involves detecting the association of particles and/or the measurement of momentum, counter-control is justified. If the total flux of particles is so high that a random expansion is likely to give tracks suitable for precise measurements (that is, for the efficient use of what magnetic field is available), then it will be too high to permit association to be appreciated, if it is low enough for such association to be distinguished, then counter-control is necessary in order to have a reasonable number of particles suitable for measurement in the field.

CHAPTER VI

THE TECHNIQUE OF PRECISION MEASUREMENTS

(1) MOMENTUM

6.1. Introduction

A particle of mass μ, momentum p, velocity βc and charge Ze (e.s.u.) moving at right angles to a uniform magnetic field, H, describes a circular trajectory of radius ρ cm., where

$$pc = \frac{\mu \beta c^2}{(1 - \beta^2)^{\frac{1}{2}}} = Ze(H\rho)$$

$$= 300Z(H\rho)\,\text{eV}. \tag{38}$$

Thus when Z is known, $(H\rho)$ is a direct measure of particle momentum.

Momentum is the most readily measured of the characteristic quantities by combinations of which the rest-mass and energy of a particle may be determined in the cloud chamber. If the mass of the particle is known, only one such quantity is required to fix its energy; if it is not, at least two quantities are needed to determine mass and energy. In either event momentum will probably be measured.

We shall discuss in detail, and with particular reference to the measurement of the high momenta encountered in cosmic-ray investigations, the factors which limit the accuracy of momentum measurement, and summarize the practical limitations arising over the whole accessible momentum scale.

Limitations on the accuracy of measurement

The relevant factors fall into three groups:

(1) The actual path followed by the fast particle through the gas of the chamber is deflected from the ideal circular trajectory, characteristic of the particle momentum, by scattering in the Coulomb field of the nuclei of atoms traversed in the gas.

(2) The photograph of drops condensed on ions formed along

this path does not accurately reproduce the actual particle trajectory. This is so

    (2.1) in counter-controlled photographs, because the ions have diffused from the line of formation, which can then only be deduced to a limited accuracy from the distribution of drops,

    (2.2) because in the time interval between the passage of the particle and the photographic exposure the drops condensed on ions defining the trajectory partake of the motion of the general body of gas in the chamber, and

    (2.3) because of distortions introduced in the optical system of the chamber and camera, and in the photographic emulsion during processing.

(3) Errors may be introduced in the procedure of measurement of curvature of the photographic image.

We consider the limitations under these headings separately, but observe as a preliminary that techniques of measurement (3), have been developed which are under all circumstances accurate compared with the distorting effects (1) and (2); that, according to conditions, either (1) or (2) may be the main source of uncertainty, and that as regards (2), (2.1) is, in foreseeable developments of technique, of small importance compared with (2.2), while the distortions (2.3) are largely systematic, with the non-systematic residues certainly also small compared with (2.2).

6.2. The scattering of cloud tracks

The scattering of fast charged particles traversing matter has been treated in detail by E. J. Williams (1939b, 1940), who gives for the mean projected angle of scattering, $\bar{\theta}$, on a plane containing the undisturbed trajectory, the relation

$$\left.\begin{aligned} \bar{\theta} &= \frac{2Ze(Nt)^{\frac{1}{2}}}{\beta(H\rho)}\,\bar{\alpha}, \\ \bar{\alpha} &= 1\cdot45 + 0\cdot80\left(7\cdot45 + 2\cdot3\log_{10}\frac{Z^{\frac{1}{3}}\sigma t}{A\beta^2}\right)^{\frac{1}{2}}, \end{aligned}\right\} \qquad (39)$$

where a thickness t of material of density σ, N atoms/cm.^{3} of atomic number Z and atomic weight A, is traversed by a fast particle, charge e, velocity βc, and momentum $300H\rho$ eV./c.

This is the arithmetic mean of the Gaussian of multiple scattering, and excludes single scattering through large angles. Single scattering from whatever cause is statistically insignificant, but in the cloud chamber there is also always a certain angle of scattering which will be recognized and which will be adopted as a criterion of rejection.

It is at first sight surprising that the random scattering of a particle passing through the cloud chamber cannot be distinguished from the regular curvature arising from magnetic deflexion, but, more closely, it is apparent that what alone can be detected is variation of the scattering curvature, and there are no grounds for connecting detectable variations with a large mean scattering curvature. That a mean scattering curvature is included in every cloud-chamber measurement of curvature was first clearly pointed out by Williams, who made a comparison of the scattering curvature with that arising in a magnetic field. More recently, Bethe (1946) has given an alternative derivation of the mean curvature which applies rather more closely to the practical procedure of curvature measurement. Bethe's criterion leads to a result similar to that of Williams but somewhat more stringent.

6.3. Treatment of track scattering by Williams

We write for $\bar{\theta}$, the mean projected deflexion in passing through a layer t of gas in the chamber,

$$\bar{\theta} = \frac{2Ze(Nt)^{\frac{1}{2}}}{\beta(H\rho)}\bar{\alpha}. \tag{40}$$

Then the uniform radius of curvature, $\bar{\rho}_s$, which would give rise to the deflexion $\bar{\theta}$, in a distance t, is given by

$$\bar{\rho}_s = \frac{t}{\bar{\theta}} = \frac{\beta(H\rho)t^{\frac{1}{2}}}{2Ze\bar{\alpha}N^{\frac{1}{2}}}, \tag{41}$$

and so

$$\frac{\rho}{\bar{\rho}_s} = \frac{2Ze\bar{\alpha}N^{\frac{1}{2}}}{\beta Ht^{\frac{1}{2}}}. \tag{42}$$

The equation (42) applies as it stands to fast particles which suffer no serious change of energy in traversing the cloud chamber. For a track 10 cm. long in air at N.T.P.

$$\frac{\rho}{\bar{\rho}_s} = \frac{40}{\beta H},$$

and corresponding expressions follow immediately for other gases and pressures. As an extreme example, in a high-pressure chamber containing argon at 100 atmospheres, the corresponding expression for 10 cm. of track is

$$\frac{\rho}{\rho_s} = \frac{700}{\beta H}.$$

Thus, in such a chamber the uncertainty of momentum measurement from this cause alone can never be less than about 5 % even for 'fast' tracks, and since, in fact, very high fields have not been used under these conditions, the uncertainties have hitherto been much larger.

A rather different relation is of importance if tracks are considered near to the end of their range, when the velocity changes from point to point in the cloud chamber. For this last part of the range of a track Williams takes a power law $R \sim v_0^n$, where v_0 is the velocity at residual range R, and n, deduced from the Bloch relation, is $\{4 - 1/\ln(137\beta_0 Z^{-\frac{1}{3}})\}$. In this approximation, he shows that

$$\frac{\bar{\rho}_s}{R} = \left(\frac{\mu}{m}\right)^{\frac{1}{2}} Z^{-\frac{1}{2}} F^{-1}, \tag{43}$$

where μ is the mass of the scattered particle, and F is a function of t, the thickness of layer over which curvature is determined and of R, the range at the beginning of the layer, but which is independent of μ and varies slowly with particle charge and velocity.

For $0.05c < v_0 < 0.5c$ and $t/R = \frac{1}{2}$, $F \sim 0.8$, and then

$$\frac{\bar{\rho}_s}{R} \sim 1.3 \left(\frac{\mu}{mZ}\right)^{\frac{1}{2}}.$$

It will be observed that the *shape* described by the ratio $\bar{\rho}_s/R$ is independent of the exact residual range and of the pressure in a particular gas, and is characteristic only of the scattering material and the mass of the scattered particle.* In air, $\bar{\rho}_s/R \sim 0.5$ for electrons, ~ 7 for mesons, ~ 20 for protons and ~ 40 for α-particles. These ratios are thus distinctive properties of the particles listed, and so may be used in the absence of field as recognition features; they also allow direct comparison with field curvature.

---

* This similarity of shape will extend also to the tracks formed directly by particles in a photographic emulsion.

6.4. Modified treatment by Bethe

Recently, Bethe (1946) has used a rather different criterion of the effective curvature of a scattered track. Consider the projection P_1MP_2 of the track on a plane, and Q_1MQ_2, the tangent through M, the mid-point of the track projection (fig. 18), and write

$$P_1Q_1 = y_1, \quad P_2Q_2 = y_2, \quad Q_1M = MQ_2 = x.$$

Then, following Williams,

$$\overline{y_1^2} = \overline{y_2^2} = \tfrac{1}{3}x^2\overline{\theta^2},$$

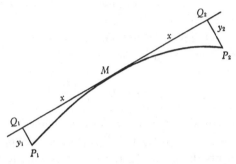

Fig. 18.

where $\overline{\theta^2}$ is the mean square projected angle of scattering in a thickness x, of the scattering gas. The sagitta of the curve P_1MP_2, s, given by $s = \tfrac{1}{2}(y_1+y_2)$, clearly represents closely the quantity which determines the estimated curvature of the track. Now y_1, y_2 are statistically independent, and so

$$\left.\begin{aligned}
\overline{(s^2)} &= \tfrac{1}{4}\overline{(y_1^2+y_2^2)} = \tfrac{1}{6}x^2\overline{\theta^2}, \\
\overline{\rho_s} &= (\tfrac{3}{2})^{\frac{1}{2}} x(\overline{\theta^2})^{-\frac{1}{2}}.
\end{aligned}\right\} \tag{44}$$

In this expression for $\overline{\rho_s}$, Bethe uses the mean square deflexion

$$\overline{\theta^2} = \frac{4\pi e^4 ZNx}{p^2\beta^2c^2}\ln\frac{\theta_{\text{max.}}}{\theta_{\text{min.}}}, \tag{45}$$

where scattering takes place in thickness x of material N atoms/cm.[3] of atomic number Z, of a particle of momentum p, and velocity βc; and where the scattering limits, $\theta_{\text{min.}}$, $\theta_{\text{max.}}$, are set respectively, by screening,

$$\theta_{\text{min.}} = \frac{mc}{p}(Z^{\frac{1}{3}}/181),$$

and by the angle, $\theta_{\text{max.}}$, which is readily recognized for a single deflexion, and is probably of the order $0 \cdot 1$ radian. Adopting this maximum, to small changes of which $\overline{(\theta^2)}$ is clearly insensitive,

$$\bar{\rho}_s = 103 \frac{\mu \beta^2}{mZ} \left(\frac{x}{BP} \right)^{\frac{1}{4}} \text{cm.}, \tag{46}$$

where B is a slowly varying factor,

$$B = 1 + 0 \cdot 44 \log_{10} \left(\frac{p\theta_{\text{max.}}}{mcZ^{\frac{1}{4}}} \right),$$

and $N = PL$, where L (the Loschmidt number) $= 2 \cdot 7 \times 10^{19}$, and P is the number of nuclei per molecule.

The relation of this scattering curvature to that in a magnetic field, H, may be expressed in the form

$$\frac{\bar{\rho}_s}{\rho} = \frac{\beta}{\beta_0}, \quad \beta_0 = 16 \cdot 5 \frac{Z}{H} \left(\frac{BP}{x} \right)^{\frac{1}{4}}, \tag{47}$$

where the velocity $\beta_0 c$ represents a critical value below which the major contribution to curvature arises from scattering, and where β_0 also represents the probable uncertainty of curvature measurement on account of scattering for fast particles ($\beta = 1$). Values of β_0 for gases at N.T.P. are given in Table XIV, and the probable error from scattering of curvature measurements on fast particles shown diagrammatically in fig. 23.

TABLE XIV. *Values of the critical velocity, $\beta_0 c$, at N.T.P.*

Gas	$H\beta_0 x^{\frac{1}{4}}$	β_0 ($H = 10^3$ gauss, $x = 10$ cm.)
H_2	25	0·008
He	35	0·011
N_2	176	0·056
A	310	0·098

This treatment leads to curvature uncertainties distinctly greater than those estimated by Williams. Thus for 20 cm. of track in air at N.T.P. and $\beta = 1$, Williams gives $\rho/\bar{\rho}_s = 28/H$ and Bethe $\rho/\bar{\rho}_s = 56/H$. The difference arises from a number of causes. Williams's result refers to the arithmetic mean deflexion, whereas Bethe uses the mean square deflexion; the former therefore gives a radius greater by a factor $(\frac{1}{2}\pi)^{\frac{1}{2}}$. Moreover, Williams refers curvature directly to the angle between tangents at the extremities of a segment of track, while Bethe considers that the directions of tangents at these points are not the features which are effectively appreciated in practical measurements, and uses instead the lateral displacement of the centre of a track segment relative to the chord. Bethe's criterion here appears to correspond more closely to the practice of measurement; it leads to a mean radius due to scattering which is again smaller than that given by Williams's treatment, in this case by a factor $(\frac{3}{4})^{\frac{1}{2}}$. (It may be observed that Bethe's criterion is still somewhat crude when, as will be shown in a later section (§ 6.12) the precision of measurement now employed requires that the measured track curvature be referred to at least a large fraction of all the drop images in the track, and so to points over the whole length of track. The development of a criterion which refers to this procedure of measurement is a matter of some importance.*) The remainder of the difference arises from the treatment of the limiting values of the impact parameter of scattering, in particular the lower limit. Bethe considers a limit which may in some cases be set by a recognizable single deflexion of the particle trajectory, which would not in practice be included as part of an attempted determination of curvature.

We postpone a discussion of the conditions under which track scattering represents the main uncertainty of curvature measurement until after a consideration of chamber distortion.

6.5. The distortion of track images

The scattering discussed in the preceding paragraphs leads to uncertainties of track curvature which may be unambiguously described in terms of the composition of the gas in the cloud chamber. The uncertainties which we group under the heading 'Chamber distortions' are of an essentially different nature. These may be considered to arise from the day-to-day accidents of a particular apparatus, and although the causes of disturbance can be studied and minimized, what remains (and in much work this will still be the major error) is quite individual to the apparatus.

* This was pointed out to me in conversation by Professor Blackett.

We are therefore concerned,

(1) with measures to reduce chamber distortions to a minimum, and

(2) with the measurement and description of these distortions, so that the meaning of curvature measurements in given circumstances can be assessed.

It is not possible for a critical person, not familiar with the particular apparatus, to form an adequate judgement of the precision of measurements from general experience, and so when any degree of precision is claimed for curvature measurements, the evidence describing the extent and variation of track distortion should be given.

We consider separately the distortions due to bulk movement of gas in the chamber and distortions in the optical and photographic system.

6.6. Gas motion in the cloud chamber

If the gas of the chamber is in motion at the instant of expansion, this motion will not be appreciably damped in the time between expansion and photography, and so will lead to distortions which will be linear in time measured from the passage of the fast particle. At expansion, three further motions may be distinguished:

(1) the actual motion of expansion; this is intended to be dead-beat, and a pure magnification in one (normal chamber) or two (radial-motion chamber) dimensions;

(2) the induced convection arising from the heating of the adiabatically cooled chamber gas by the uncooled walls;

(3) localized irregularities of motion, associated particularly with the motion of rubber diaphragms, and leading to turbulent 'puffs' in the chamber.

With regard to (3), these local irregularities are mainly of importance because of the associated condensation which is often present. If the velocity of propagation is large enough for the disturbance to reach the central part of the chamber, the motion will also be sufficient to give recognizable deformation of tracks. A striking example was given in the classical paper by Blackett and Occhialini (1933) (Plate 24, photograph 15).

The other factors lead to distortions which are functions of time and of position in the chamber, and which are only easily detected by inspection in gross examples. The development of distortion with time sets a limit to the tolerable interval between the passage of a particle and photography (§ 5.2), and the variation of distortion over the chamber may make necessary a correcting map over the parts of the chamber used.

In practice, it is found that almost all chambers may be regarded as dead-beat, and hence the act of expansion leads to distortion which is independent of time. In contrast, as already stated, the pre-existing movements of the gas lead to distortions which are linear in time, while the induced convection following expansion develops with a rather high power of time. (Approximately, the total buoyancy of gas near the chamber wall increases as $t^{\frac{1}{2}}$, and so the displacement at any point in the chamber, during the initial stages of motion, will be proportional to $t^{\frac{5}{2}}$.) However, the induced convection does not begin until expansion is in progress, and it is found to be straightforward to keep the time elapsing before photography short enough to reduce distortions from induced convection to secondary importance.

Pre-existing gas motions

The pre-existing gas motions are probably the most important source of varying distortion. These arise because of temperature differences within the chamber, that is, between different parts of the chamber wall, and are particularly conspicuous at the surface of an experimental plate inside the chamber; partly because such a plate is thermally rather isolated, tending to lag behind the outer walls in changes of temperature; and partly because the motion is most conspicuous near to the surfaces at which motion is initiated (where there is the maximum normal velocity gradient), and this is the only surface in the chamber close to which it is necessary to study tracks. A typical example of rather severe distortion at a plate is shown in fig. 19, together with an indication of the gas motion concerned; the example refers to a chamber with a heated lamp, clearly not thoroughly thermally screened from the chamber, at the right-hand side.

External heat sources to which the chamber may be subjected include the exciting coils for magnetic field, together with the cooling system of the coils if this is not stabilized, and the lamp for photographic illumination if this is of a preheated type. There is a strong case for building the whole cloud chamber into a box of high conductivity, the temperature of which is stabilized thermostatically. Such a box would include the lamps, and these would therefore have to be of an unheated type. The complete solution does not appear so far to have been applied; Blackett and the writer (1937), working with a magnet of very large yoke weight ($\sim$ 10 tons),

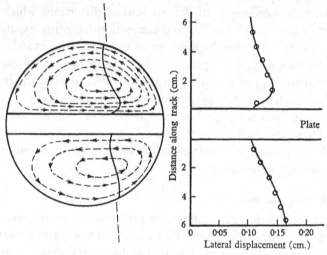

Fig. 19. Distortion of a counter-controlled cloud track by convective drift. Left, schematic; right, measurements on an actual track.

aimed at maintaining the cloud chamber at the temperature of the magnet yoke, surrounding all parts not in direct contact with the iron in a fibre-board enclosure, and admitting light from a carefully cooled lamp through a thick plate-glass window, and this partial solution was conspicuously successful; more recent developments have involved the stabilization of temperature of the room containing chamber and magnet (Rochester and Butler, 1947).

The gas in a chamber at completely uniform temperature is in neutral stability, and slight temperature variations can initiate motion. By cooling very slightly the lowest part of the chamber, the gas can be brought into positive stability; it will settle down more quickly after compression and will be less sensitive to slight temperature disturbances. This cooling may be done by means of a very slow water flow through a metal pad on the outer surface of the chamber wall at the bottom, or in a chamber constructed in metal or in Perspex the water channel can be cut directly in the chamber wall. As has already been pointed out (§ 4.3) the drawback of imposing positive stability on the gas of the chamber arises because convective motion after compression is an important mechanism for restoring saturation to the gas as it warms up. If this convection is suppressed prematurely, saturation in the upper parts of the chamber must be completed by diffusion and a relatively long time must be allowed for this process.

In a chamber without complete control of temperature, the distortions are likely to show a strong systematic element; for example, in the situation sketched in fig. 19, the presence of a preheated lamp is likely to maintain permanently a motion of the type indicated, although the system may require many hours to reach a stable condition. As control of temperature is improved, the nature of any disturbance becomes less strongly systematic, and experience suggests that a stage may be reached where systematic distortion from these causes is, throughout the chamber, small compared with the random spread. When this stage is reached, unless other factors require it, the need for a map of distortions over the chamber will be slight.

An estimate may be made of the velocities of gas motion, present before expansion, which correspond to the uncertainties of accurate curvature measurements. We take, as a standard of performance which has not been appreciably exceeded, measurements by the writer (1938) with plates of various metals in a cloud chamber of about 25 cm. effective diameter. The actual length of track measured was about 10 cm., the track inclination to the vertical did not exceed about 20°, and the uncertainty of curvature was, uniformly over several long runs, about 0·03 m.$^{-1}$, corresponding to a displacement of the centre of a track from the chord of 0·004 cm. The interval from the passage of the particle to photography was probably about 0·03 sec., and so the average difference of velocity of drift, normal to the track, of the centre compared with the ends which would account for the whole of this uncertainty of measurement is about 0·1 cm./sec. Since the ends of the tracks in this work are approaching surfaces at which motion was likely to be comparatively large, this would probably correspond to a maximum velocity of gas drift in the used parts of the chamber appreciably less than 1 cm./sec.

Distortions at expansion

The gas velocity at the surface of non-porous walls at expansion is zero, and hence at walls parallel to gas motion there is a transition layer, in which the total motion at expansion is less than that in the free parts of the chamber. For most practical purposes this layer is much less than 1 cm. thick and is of little importance, but it offers an obvious objection to making the expansion in a thin plate-like chamber in any direction but that of the small dimension.

The motions throughout the chamber may be modified by the detailed behaviour of the expansion mechanism; the piston of the chamber may not move uniformly but may tilt during expansion, or the porous diaphragm may differ in porosity from one area to another. Such features will lead to systematic variations of distortion from point to point in the cloud chamber, and can only be detected on a distortion map of the chamber (see §6.8 below).

(*Experience in the laboratory of Professor R. B. Brode with a chamber under very critical temperature control shows that significant distortions can arise from residual oscillations of the outflowing gas from the back of the chamber. The chamber was of the type in which a rubber diaphragm moves between defined positions, and it was found that distortion in the cloud chamber was reduced when the outflow of air was damped by a succession of porous diaphragms, in spite of the resulting reduction in speed of expansion.)

6.7. Distortion of the optical system

Normal photographic lenses show residual distortions of field which, while insignificant for most purposes, are of importance in the measurement of track curvature. These distortions lead to spurious curvature of tracks which do not pass through the axis of the lens which may be expressed in terms of the angular distance of the nearest point of the track from the axis.

* Added November 1949. I am indebted to Professor Brode for a discussion on this problem.

It is possible, by special correction in manufacture (Blackett, 1937), very much to reduce the lens distortion, and in fig. 20 the curvature introduced by the ordinary version of a typical lens is compared with that given by a specially assembled lens. The scale of curvatures in fig. 20 refers to the image space, and, under operating conditions, of magnification 0·1 and magnetic field 10^4 gauss, a curvature of 1·0 m.$^{-1}$ on this scale is equal to the field deflexion of a particle of momentum 3×10^9 eV./c.

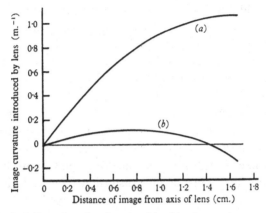

Fig. 20. Optical distortion of a photographic objective used at a magnification 1/5·6. Curve (a), normal lens; (b) specially assembled lens. (Blackett.)

The front glass plate of a cloud chamber may vary from $\frac{1}{4}$ to 1 in. in thickness, and a further spurious curvature is introduced when tracks are viewed obliquely through such a plate. However, since the angle of viewing does not as a rule exceed 20 or 30°, and the plate thickness is a small fraction of the object distance, the magnitude of curvature introduced is likely to be small compared with that due to the lens.

6.8. Treatment of chamber distortions

We shall consider briefly the way in which the systematic element in chamber distortion is treated. It may be advisable to use distortion-corrected lenses, and it may be necessary to make a map of systematic distortions covering the part of the chamber used. The use of a map of distortions is probably only feasible when the illuminated part of the chamber is only a small fraction of its depth, well separated both from the front and back walls, and when the inclination of tracks to the vertical is limited. Under these conditions a distortion map is one-dimensional, distortion curvature being expressed as a function only of the position across the chamber of the centre of the track. It will be noticed that optical curvature introduced by the photographic lens can for these conditions be represented in the same way, and so there is then no great advantage in using corrected lenses.

The distortion map is based on photographs taken in zero magnetic field (the residual field of the magnet should be checked and, if necessary, eliminated). It is probably advisable to exclude the slowest cosmic-ray particles from the zero field photographs, and to do so it will be sufficient to include 10 cm. of lead in the defining counter-telescope below the chamber, thus excluding all particles of momentum less than about 3×10^8 eV./c. The zero field photographs should be interspaced in time among those for which they are to yield corrections, and for most purposes it is not necessary to measure more than about 100 of these in all, for there is no advantage in defining the systematic distortion to a precision much greater than the residual random distortion. The crude distribution of curvatures among these tracks gives immediately the uncertainty of curvature measurements without regard to position in the chamber. The track curvatures are then plotted against the co-ordinate across the chamber of the centre of each track, and a correction curve drawn. A second distribution of the residual curvatures, that is, the departures of actual curvatures from the chosen correction curve, gives the final uncertainty of curvature measurements throughout the series of photographs.

A typical calibration plot,* which refers to a cloud chamber divided into upper and lower halves by a 2 cm. metal plate, and in which the curvature of zero field-track images is plotted against the horizontal co-ordinates of the mid-points of tracks in either half of the chamber, is given in fig. 21. In this example, particularly in the upper half of the chamber, the systematic curvature is fairly constant over the full breadth of the chamber, but the halves show marked and differing systematic curvature. The points plotted in

* The example is taken from the data of measurements described by the writer (Wilson, 1938).

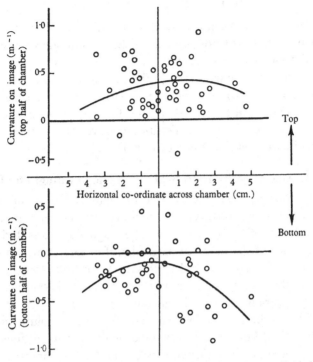

Fig. 21. Map of curvature corrections in a cloud chamber divided by a central metal plate.

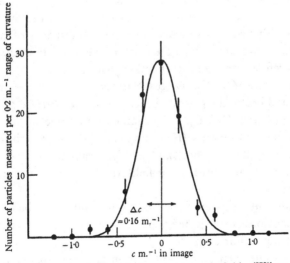

Fig. 22. Residual curvature of tracks in zero magnetic field. (Wilson, 1938.)

the diagram have been corrected for the distortions of the optical system; if this had not been done, the magnitude of distortion would have been considerably greater. The scale of curvatures refers directly to curvatures on the photographic negatives, and a curvature $C = 1 \, \text{m.}^{-1}$ on the negative here corresponds to $C = 0 \cdot 114 \, \text{m.}^{-1}$ in the chamber, and, for the magnetic field used, to $p = 2 \cdot 6 \times 10^9 \, \text{eV./c.}$ The distribution of residual random curvatures of corrected tracks is shown in fig. 22.

6.9. Check of calibration in the presence of magnetic field

The procedure outlined in the previous section is of general application but is open to objection because the calibration used when the chamber operates in a large magnetic field is derived in the absence of the field. In particular, if a significant part of the systematic distortion of tracks arises from non-uniformity of piston motion, the conditions are clearly not equivalent because of eddy-current restraint on the piston movement in the field. It is not possible to check with full magnetic field the absolute value of a calibration, but when the chamber is divided in sections some calibration differences may be checked, and for an important range of work this is of value.

The check, which is specific to cosmic-ray particles, is based on the assumption that a high-energy meson passing through a metal plate loses only a small fraction of its energy, and hence suffers a minor change of true curvature. The differences of measured curvature, on either side of the plate, of the tracks of these particles are therefore attributed to chamber distortion.

If we write C_T, C_B, $\bar{C}$, respectively for curvatures measured above and below the plate and their mean, *all corrected to the zero-field calibration*, $C_T - C_B$ is determined for all tracks with $|\bar{C}|$ less than some limiting value (for example, $|\bar{C}| < 1 \, \text{m.}^{-1}$ in the negative, $\sim 0 \cdot 1 \, \text{m.}^{-1}$ in the chamber), and any significant deviation of $C_T - C_B$ from zero then represents a field variation from the zero field calibration.

Finally, an estimate of curvature uncertainty may be made from the tracks measured in magnetic field. The data may be used in several ways. For example:

(a) Assume that all examples of an apparent change of sign of curvature in fast tracks are spurious. Then it is easily shown that the probable error of curvature, ΔC, is given by

$$\Delta C = 0{\cdot}42 \frac{\Sigma_n(|\,C_T\,| + |\,C_B\,|)}{n} \tag{48}$$

from tracks in this category.

(*b*) Assume that all examples of apparent gain of energy (without change of sign) by fast particles are spurious; a similar relation may be deduced.

It will be observed that these checks are essentially tests of internal consistency of a complete set of curvature measurements, and that these tests are not necessarily applicable to *selected* groups of tracks; for example, they are not applicable if the lower momentum particles alone have been selected for measurement.

6.10. Maximum detectable momentum

As a measure of performance of a particular apparatus, the *maximum detectable momentum*, p_0, the particle momentum for which true field curvature is equal to the probable uncertainty of curvature measurement, is a useful standard of comparison. Our knowledge of a particle measured to have this momentum or greater is obviously slight; even the sign of charge is uncertain, and we only know that the momentum is probably greater than some smaller value, say about $\frac{1}{4}p_0$. Lower measured momenta, p, have a probable uncertainty of momentum p^2/p_0 arising from distortion in the cloud chamber.

Since the performance of a particular apparatus is necessarily unique, only rough indications of standards of performance can be given. In a 30 cm. diameter chamber at about atmospheric pressure, used with particles not more than 20° from the vertical, a maximum detectable momentum of from 2 to 5×10^{10} eV./c. can probably be reached using a field of 10^4 gauss or greater. The corresponding figures for tracks in each half of the chamber when this is divided by a metal plate is from 1 to 2×10^{10} eV./c. At lower magnetic fields, since heating problems are less severe, the performance is probably rather better than that given by the ratios of field strengths.

The variation of p_0 with chamber size has not been investigated in any detail, and while the general improvement is probably not more than linear, as size increases it becomes much easier to avoid boundary layers of gas, the thickness of which is not a function of chamber size, and so the practical gain is marked. Equally, the variation of p_0 with track inclination has not been studied, although it is practically relevant to some important phenomena, notably to the investigation of the products of nuclear evaporations which may lead to an almost isotropic particle emission. It seems likely, however, that a reduction of p_0 of at least two or three times takes place if tracks in all directions in the chamber are included.

6.11. The relative importance of track scattering and chamber distortions

It will be useful to give here a summary indicating the domains over which track scattering and chamber distortions respectively are of major importance. Following Bethe (§ 6.4), $\Delta_1 p$, the probable scattering uncertainty of the measured momentum p, is given by

$$\frac{\Delta_1 p}{p} = \frac{\rho}{\bar{\rho}_s} = 16 \cdot 5 \frac{Z}{\beta H} \left(\frac{BP}{x}\right)^{\frac{1}{2}} = \frac{\beta_0}{\beta},\tag{49}$$

while that due to chamber distortions, $\Delta_2 p$, is

$$\frac{\Delta_2 p}{p} = \frac{p}{p_0}.\tag{50}$$

Then, $\Delta_1 p$, $\Delta_2 p$ are of equal importance at $p = p'$, where

$$\beta p' = \beta_0 p_0.\tag{51}$$

In this relation, p_0 is, to a minor and uncertain degree, a function of Z, P and x, and is roughly proportional to H, but these variations are secondary to the major variations arising from detailed technical handling, and we here assume as a basis of discussion typical values of p_0.

In Table XV, the critical values of $\beta p'$ below which scattering errors predominate are given for a range of gases and pressures and for typical values of p_0 which correspond to a high standard of technique in low and high magnetic fields for 20 cm. of track. Of the curves in fig. 23, (1) and (2) refer to conditions similar to those tabulated in Table XV for argon at a pressure of one atmosphere, (3) refers to a high-pressure chamber with a magnetic field of intermediate value, and (4) gives the performance reasonable for a chamber containing a metal plate and in which particles in all directions must be measured.

TABLE XV. *Typical values of βp', at which track scattering
and chamber distortion are of equal importance*

Gas	20 cm. track, $H = 10^3$ gauss, $p_0 = 5 \times 10^9$ eV./c. Pressure in atm.			20 cm. track, $H = 10^4$ gauss, $p_0 = 3 \times 10^{10}$ eV./c. Pressure in atm.		
	1	10	100	1	10	100
H_2	$4 \cdot 0 \times 10^7$	$1 \cdot 3 \times 10^8$	$4 \cdot 0 \times 10^8$	$2 \cdot 4 \times 10^7$	$0 \cdot 8 \times 10^8$	$2 \cdot 4 \times 10^8$
He	$5 \cdot 5 \times 10^7$	$1 \cdot 7 \times 10^8$	$5 \cdot 5 \times 10^8$	$3 \cdot 3 \times 10^7$	$1 \cdot 0 \times 10^8$	$3 \cdot 3 \times 10^8$
N_2	$2 \cdot 9 \times 10^8$	$0 \cdot 9 \times 10^9$	$2 \cdot 9 \times 10^9$	$1 \cdot 7 \times 10^8$	$0 \cdot 5 \times 10^9$	$1 \cdot 7 \times 10^9$
A	$4 \cdot 8 \times 10^8$	$1 \cdot 5 \times 10^9$	$4 \cdot 8 \times 10^9$	$2 \cdot 9 \times 10^8$	$0 \cdot 9 \times 10^9$	$2 \cdot 9 \times 10^9$

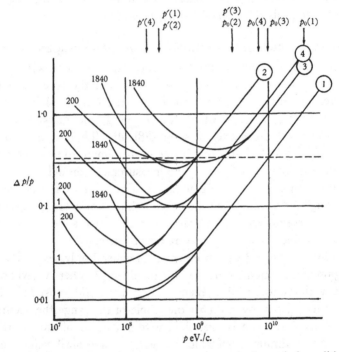

Fig. 23. Probable errors of momentum measurement for typical conditions, using Bethe's criterion of scattering.

1. 20 cm. track, $p_0 = 3 \times 10^{10}$ eV./c., $H = 10^4$ gauss, 1 atm.; argon.
2. 20 cm. track, $p_0 = 3 \times 10^9$ eV./c., $H = 10^3$ gauss, 1 atm.; argon.
3. 20 cm. track, $p_0 = 10^{10}$ eV./c., $H = 3 \times 10^3$ gauss, 100 atm., argon.
4. 10 cm. track, $p_0 = 5 \times 10^9$ eV./c., $H = 8 \times 10^3$ gauss, 1·5 atm.; argon.

Very little meaning can be attached to individual measurements when $\Delta p/p > 0 \cdot 3$ (horizontal broken line). Lines marked 1, 200, 1840 are appropriate respectively to electrons, mesons and protons. p_0 is the maximum detectable momentum, p' is the momentum at which scattering and chamber distortion are of equal importance.

When $\Delta p/p$ in the diagram is greater than about $\frac{1}{3}$, the conclusions which may be drawn from an individual track are valueless—even the sign of curvature is uncertain—although observations of a sufficiently large group may be statistically significant. The distinction between criteria appropriate to individual tracks and those significant for groups is important, because it is common for tracks which are at sight abnormal to be selected for quantitative interpretation from a much larger mass of data which is not treated in any detail. We are then concerned to know whether the properties of the selected tracks are consistent with the extreme members of the random distribution of disturbances for the whole group.

6.12. Curvature measurements on the photographic image

We have discussed in the last sections the precision with which the track image photographed will represent the true curvature corresponding to particle momentum; the numerical limits described are of a technically advanced kind, and marked further improvement is a matter of some difficulty. There is therefore a very clear standard of actual image measurement to be sought, for the errors introduced during the measurement should be insignificant compared with those already discussed. It is also desirable, although not so important, that the method of measurement should give a survey of the whole available track so that sudden deflexion or noticeable change of curvature may be detected.

This standard of measurement is in general only severe for the region of high momentum, say $p > 10^8$ eV./c., and here we will only deal with the technical problems under these conditions. It is clear from fig. 23 that over almost the whole of the range the accuracy of image formation is described within a factor of two or three by the maximum detectable momentum, a typical value being $p_0 = 2 \times 10^{10}$ eV./c. on 20 cm. track in a field $H = 10^4$ gauss, and so $\Delta C = 0.015$ m.$^{-1}$ Hence we aim at methods of measurement which will detect curvatures as little as $C = 0.005$ m.$^{-1}$ on a length of 20 cm. of track.

Now the sagitta of a 20 cm. arc of radius 2×10^4 cm. is about 0.003 cm., while the breadth of a counter-controlled track is of the order 1 mm.; thus the trajectory must be located to very much within the diffusion breadth, and to do this, apart from errors of setting, a certain minimum number of drops must be used in fixing the points from which curvature is deduced, and so a certain minimum length of track used to determine each point.

The mean projected distance of diffusion, on a plane containing the initial trajectory is (§ 5.2 above),

$$\bar{x} = 0.95 \sqrt{(D\tau)}, \tag{52}$$

where D is the coefficient of diffusion of ions, and τ the time of diffusion from formation until diffusion is stopped by condensation. If $\tau = 0.014$ sec., and D (air at N.T.P.) $= 0.034$ cm.^{2} sec.$^{-1}$,

$$\bar{x} \sim 2 \times 10^{-2} \text{ cm.},$$

as compared with the required uncertainty in determining the sagitta of a track

$$\delta s \sim 3 \times 10^{-3} \text{ cm.}$$

Hence, considering only primary ionization, for which the assumption of linear ion formation is strictly applicable, about 100 drops must be used to determine δs, and in air at N.T.P., where the primary ionization is about 20 ion pairs per cm., of which some at least will be so close to regions of secondary ionization as not to be available, at least $2\frac{1}{2}$ cm. of track must be used to determine each of *three points* on which a sagitta measurement is based. Since this estimate takes no account of the technical difficulties of making the actual setting in any measuring apparatus, the actual length needed may well be twice as long.

The corresponding figures including secondary ionization are not easily given, since there is no systematic information about the formation distance of secondary ions from the primary trajectory. The smaller blobs along tracks result by diffusion from initially very much smaller groups of ions, and so the centre of such a blob may be nearer the true trajectory than the image of a single droplet. Larger blobs, with a size at formation of the same order as subsequent diffusion distances, tend to be appreciably off the true trajectory. While use will undoubtedly be made of the smallest secondary clusters, the appearance of the clusters in a photograph is so variable that all but the smallest must be discarded to ensure that displaced groups are not included. The length of track including small secondary clusters required to define each point is probably somewhat, but not greatly, smaller than that required using only distinct primary droplets.

The estimates given refer to air at atmospheric pressure; in argon at $1\frac{1}{2}$ atmospheres, a rather shorter length of track will be needed.

There is clearly serious objection to the simple, three-point determination of curvature, quite apart from the failure of such a method to give any indication of abnormal distortions, for unless each point is based on considerable track length, the method is in principle not good enough to use the full accuracy of the photographs. The two methods which are in fact used are based on observations covering at least the greater part of each track, and these are described below.

The statistical spread of droplets imposes a particular limit on the measurement of very strong curvature, when the radius of curvature is, say, 1 cm. or less. Then a co-ordinate setting on the circumference is limited to a precision defined by that length, $2l$, of arc in which the displacement of the arc from the tangent is of the same order as the precision of setting allowed by the total number of drops in the length $2l$. If the radius is ρ, the precision of radial setting $\delta\rho$, the semi-breadth of the track, x_0, and the primary drop density N_0 per cm. of track, then

$$\delta\rho \sim \frac{x_0}{\sqrt{(2N_0l)}}, \quad 2\rho\delta\rho = l^2,$$

and so

$$(\delta\rho)^5 \sim \frac{x_0^4}{8N_0^2\rho}. \tag{53}$$

If $x_0 = 0.05$ cm., $N_0 = 20$ cm.$^{-1}$, the following relations are obtained between ρ and $\delta\rho$ (Table XVI).

TABLE XVI

ρ (cm.)	$\delta\rho$ (cm.)	$\delta\rho/\rho$
1·0	0·018	0·018
0·5	0·021	0·042
0·2	0·025	0·125
0·1	0·029	0·29

6.13. Co-ordinate plot of track image

In a method first described by Anderson (1933), a travelling microscope, traversing in two directions at right angles, is used to give a plot on a suitably modified scale of the particle trajectory, and the curvature of this is determined by the sagitta method from the smoothed curve through the points. It is convenient to use a normal travelling microscope of good quality to traverse in a direction approximately parallel to the track image and an eyepiece micrometer (a convenient scale is 1 scale div. ≡ 0·0005 cm.) to measure distance perpendicular to this direction. The micrometer should be fitted with a cross-hair parallel to the track, which is brought to the best possible fit with the whole of the track visible in the field of the eyepiece, and the intervals plotted along the track should not

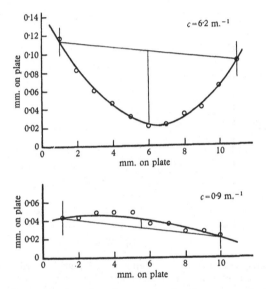

Fig. 24. Determination of track curvature by direct co-ordinate plot.

be greater than the diameter of the field of view. In a good image, attention will probably be confined as far as possible to primary ionization; under less favourable conditions secondary clusters must be used, and then advantage should be taken of the fact that the smaller the cluster the more closely is it likely to be centred on the true trajectory. A normal plot made in this way is shown in fig. 24, where the vertical (eyepiece) scale is magnified relatively to the horizontal fifty times.

An alternative method of plotting in which the differences of micrometer readings for equal steps along the track are plotted against co-ordinates along the track may be used provided that the total deviation of the image is not too great; to the accuracy we have been considering, not greater than about 8° of arc. Although this method of plotting (fig. 25) has little advantage for most purposes, it brings out very clearly the departures of individual readings from the line finally adopted. Using apparatus with approximately the performance stated, the probable distance of an individual setting from the described trajectory was found to be about 0·0004 cm. on

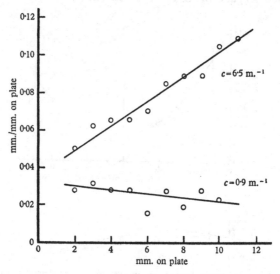

Fig. 25. Determination of track curvature by differential co-ordinate plot. (The track measurements illustrated here are the same as those used in fig. 24.)

the photographic plate for a good track and about 0·0006 cm. for an image of moderate quality. The photographic magnification was about 1/10, the cloud chamber contained argon at 1½ atmospheres pressure, and the field of view of the microscope covered rather less than 2 cm. of track in the chamber; the precision of setting therefore agrees well with that estimated.

The raw data obtained in this way is probably good enough to meet the criterion laid down in §6.12, but its rigorous use is too tedious for application except for tracks of exceptional interest. In the quick alternative method, in which the sagitta of a freely drawn curve is measured, the accuracy is probably distinctly lower.

6.14. Optical compensation of track curvature

A most elegant null method in which track curvature is compensated by a known superposed curvature is due to Blackett (1937), and was improved in detail by Ehrenfest. It is very much quicker in use than co-ordinate plotting at the same level of accuracy, although it is probably no more accurate in the limit.

When a track image of uniform curvature is correctly compensated the compensated image is a straight line, and the striking ability of the eye to appreciate deviations from straightness is used

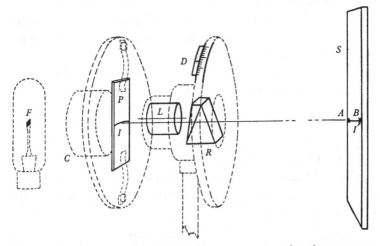

Fig. 26. Prism curvature compensator for measurements of track curvature.

to establish the null point; this ability to judge straightness is used to the full if the final image, on a flat screen, is viewed, much foreshortened, at almost grazing incidence.

The method has considerable advantage in principle. The whole track image is in view at once, and displacements of individual clusters from the trajectory line are shown up in track images which cannot be made straight by a uniform compensating curvature.

Of several optical devices for introducing a known, calibrated curvature, that which has been widely used is an achromatic prism rotated about an axis perpendicular to the refracting edge, and the apparatus is shown diagrammatically in fig. 26. The projection system (lamp, F, condenser C, lens L) throws an image I', of the track I, on the screen S, through the achromatic prism R; the lens and the prism are mounted together on a graduated head rotating about the axis through the centre of I; the position of the head is read at the scale D. The axis of the lens, L, is parallel to that of rotation but may be displaced (as shown in the diagram) by the amount required to compensate the total deflexion of the prism. This refinement is not essential but serves to keep the image, I', approximately stationary as the head carrying L and R is rotated. If this is not done, the observer, looking along the image, I', at a very small angle, must continuously alter the height of his eye as the prism head is turned, and the movement is most tiring. The screen, S, should be flat with a highly diffusing surface. A magnesium oxide surface, formed by holding the screen face downwards while magnesium ribbon is burned below, is particularly suitable; the surface comes away when touched and so cannot be cleaned, but it is so easily renewed that this is not a serious drawback.

Performance of the optical compensating instrument

The theory of curvature introduced by a prism is rather complicated, but at minimum deviation the curvature C of the image of a straight line parallel to the refracting edge of a prism of angle 2α and refractive index μ, has been given by Bravais (Blackett 1937) and can be written

$$C = \frac{2(\mu^2 - 1)\sin\alpha}{\mu f(1 - \mu^2 \sin^2\alpha)^{\frac{1}{2}}}, \qquad (54)$$

where f is taken as the distance from object plate to lens. For an achromatic prism the total curvature is approximately the difference of the components, but since it is not possible to use both components at minimum deviation the relation is not exact.

The range of curvature which can be covered by compensation of the prism type is illustrated by three instruments which have been used at Manchester. These use achromatic prisms of deviation about $9°$, and since test circles show no detectable departure from straightness when fully compensated, the limit of the method, set by the intrusion of higher orders of distortion of the compensating system, has not been reached in these instruments. The two prisms, by Beck, are of almost identical deviation, but the second introduces markedly larger curvature.

Instrument 1. Beck prism I; object-image distance $(u+v) = 39$ cm.; T.T.H. 35 mm., $f/2\cdot5$ lens mounted axially; using 1 cm. of object track. Calibration: $C = 4\cdot12 \sin \theta$ (C m.$^{-1}$ = curvature on negative).

Instrument 2. Beck prism II; $(u+v) = 46$ cm.; T.T.H. 35 mm., $f/2\cdot5$ lens mounted eccentrically to eliminate image shift, using 1 cm. of object track. Calibration: $C = 6\cdot11 \sin \theta + 1\cdot12 \sin^2 \theta$.

Instrument 3. Beck prism II; $(u+v) = 26$ cm.; T.T.H. 15 mm., $f/2$ lens mounted axially, using 7 mm. length of object track. Calibration: $C = 13\cdot6 \sin \theta$.

The third instrument can readily be used up to $C = 12\cdot5$ m.$^{-1}$. Then, if the photographic magnification is $0\cdot1$ and the magnetic field 10^4 gauss, the low momentum limit of the instrument is at $p = 2\cdot4 \times 10^8$ eV./c. It will be noted that when the projection lens is mounted eccentrically to eliminate the lateral shift of the final image (Instrument II), appreciable deviations may occur from the elementary sine calibration.

Procedure of measurement

The prism head is brought to zero setting and the plate P (fig. 26) inserted and adjusted until the image I' passes through the two reference marks A and B on the screen S. This location of I' is necessary because the distance of the track from the axis of the projection lens, and the orientation of the track relative to the zero setting of the prism head, must both be fixed. Several compensator settings which are judged to make the image I' straight are taken; the object track is then rotated through 180° and a second set of readings taken. The curvature is determined from the mean value of the two settings.

The two settings, with the uncompensated track respectively concave upwards and convex upwards, frequently differ by two or three degrees, and similar discrepancies occur throughout a set of tracks measured in a single session at the instrument. It appears that the personal standard of straightness of the individual observer is not absolute but shows deviations from true straightness which vary only rather slowly with time. It is likely that this standard changes little between the two halves of the observation on a single track, and so the deviation from a true standard is largely eliminated by reversing the track in the manner described.

Precision of setting

Examination of sets of measurements with the compensator allows an estimate of precision to be made; the figures quoted here refer to the instrument (no. II) described above.

The mean compensator readings for the two track orientations are θ and θ', the adopted compensating angle is then $\frac{1}{2}(\theta - \theta')$ (θ and θ' will in general be of opposite sign). We may estimate the precision of setting by noting the deviations of individual settings from θ or θ' as the case may be, deducing from these deviations the uncertainty of the means, θ and θ'. Alternatively, we may try to deduce directly the uncertainty of the mean, for if the mean value of $\frac{1}{2}(\theta + \theta')$ differs from the predetermined zero because the observer has a non-absolute standard of straightness, this should also persist between successive tracks measured (in particular, when there are two tracks, above and below a plate, in a single negative). If the two tracks are distinguished by subscripts 1 and 2, the quantity

$$\{(\theta + \theta')_1 - (\theta + \theta')_2\}$$

should depend only on the precision of setting.

Over a routine set of measurements, individual readings for tracks with $C < 1$ (m.$^{-1}$ on the plate) showed a mean deviation $\Delta C \sim 0.1$ from the appropriate mean. If $\frac{1}{2}(\theta - \theta')$ were in each case based on eight such settings, the expected uncertainty of the final curvature would be $\Delta C \sim 0.035$. For the same group of tracks, however, the mean value of $\{(\theta + \theta')_1 - (\theta + \theta')_2\}$ corresponds to a curvature deviation $\Delta C \sim 0.32$ and the corresponding uncertainty of the adopted curvature, $\frac{1}{2}(\theta - \theta')$, is $\Delta C \sim 0.11$.

The actual precision of measurement probably lies between these values; that from individual settings ($\Delta C \sim 0.035$) is probably low, since there is a tendency to repeat any set of readings which is not concordant to some preconceived standard; that from differences of mean readings ($\Delta C \sim 0.11$) is certainly high, because it cannot be completely independent of differing judgements of the appreciable non-circular elements of chamber distortion. It is reasonable to estimate that the precision of measurement under routine conditions lies between $\Delta C = 0.04$ and $\Delta C = 0.10$ on the plate for about 1 cm. of track image, which substantially reaches the standard of measurement laid down at the beginning of §6.12 for a magnification in the photograph of about 0.1.

(2) ANGULAR MEASUREMENTS

We shall discuss here in the first place the general problem of the measurement of the angle between straight tracks in the cloud chamber. Secondly, we consider the special problems, analogous to those of momentum measurement, which arise in connexion with the small angles of scattering of cosmic-ray tracks photographed in a counter-controlled chamber.

6.15. General features of angular measurement

The three-dimensional orientation of tracks is obtained by working back from two photographs. If these are taken in planes at right angles, the reduction from the measured angles on the plate may be made analytically, and such a disposition will be used to obtain maximum precision of measurement. If less precise measurements are required, or if the possible orientations of tracks are limited to within small angular distances from a fixed plane, the relative positions of photography may be chosen more freely, and the lenses may be placed as a stereoscopic pair, and the advantage of stereoscopic viewing combined with the provision of photographs for reprojection. Reprojection methods suffer from the drawback that the lens and the camera assemblies used for reprojection must be exact replicas of those used to take the photographs, and in practice the only satisfactory procedure is to use this part of the original camera for projection.

6.16. Photographs at right angles

In the arrangement developed by Shimizu (1921) (fig. 27), a system of mirrors is used to bring two photographs, taken effectively at right angles, on to a single plate using a single lens. The method suffers, however, since the focal planes of the two views are at right angles, and so only a very small chamber volume is in focus in each of the two views. Since for the lens apertures which are essential for cloud-chamber work the volume which is in focus is always a flat plate, this arrangement is inefficient, even if the orientation of tracks to be measured is quite unrestricted, unless the position in space of the tracks is fixed. In his study of the nuclear collisions of α-particles and of disintegrations produced by them, Blackett (1922, 1923, 1925) therefore developed a modified arrangement in which a flat plate-like volume was made to be the in-focus

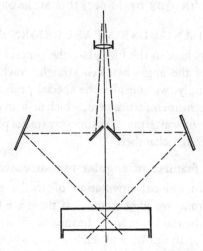

Fig. 27. Photography of tracks in directions at right angles. (Shimizu.)

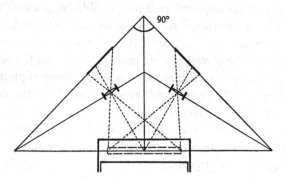

Fig. 28. Photography of tracks on planes at right angles. (Blackett.)

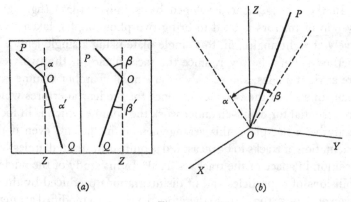

(a)   (b)

Fig. 29.

volume of two photographs at right angles (fig. 28). The analytical treatment of photographs in this system is identical with that for Shimizu's method.

In a first approximation, the two photographs are rectangular projections of the track system on planes at right angles. Fig. 29(a) represents the records of two copunctual tracks OP, OQ, where the dotted lines, OZ, represent the line through O parallel to the intersection of the planes of projection. If the projection is rectangular, the photographs thus represent the projection of the lines OP and OQ (OQ not shown) (fig. 29(b)) on the planes XOZ, YOZ. Then if

$$OP \text{ is } l, m, n,$$
$$OQ \text{ is } l', m', n'.$$

The angle $POQ = \phi$ is given by

$$\cos \phi = ll' + mm' + nn',$$

and using the four relations like $l/m = \tan \alpha$,

$$\cos^2 \phi = \frac{(1 + \tan \alpha \tan \alpha' + \tan \beta \tan \beta')^2}{(1 + \tan^2 \alpha + \tan^2 \beta)(1 + \tan^2 \alpha' + \tan^2 \beta')}. \tag{55}$$

This expression reduces to simple forms in the special cases

$$\alpha', \beta' = 0, \quad \tan^2 \phi = \tan^2 \alpha + \tan^2 \beta,$$
$$\alpha', \beta', \alpha, \beta \text{ small}, \quad \tan^2 \phi = \tan^2 (\alpha - \alpha') + \tan^2 (\beta - \beta'),$$

the latter being adequate for small α' and β' up to α, $\beta \sim \frac{1}{4}\pi$. Equation (55) is indeterminate when α and β (or α' and β') are equal to $\frac{1}{2}\pi$, but if $OP(OQ)$ ends in the gas, or if the track has distinguishing features such as the secondary clusters of electron tracks, a straightforward extension of the derivation to make use of a definite range of OP allows ϕ to be determined.

More exactly, the projection involved in photographic image formation is not *rectangular* but *conical*; we shall write a, a', b, b' for the actual angles in the photographs of which the values in rectangular projection would have been α, α', β, β', and determine corrections in the form $(\tan \alpha - \tan a)$, etc. The correction will be zero on the lens axis and will increase with distance from the axis. In the Blackett arrangement (fig. 28) the lens is inclined and these corrections may be considerable.

In fig. 30, $OXYZ$ defines the object space, and $X'OZ'$, $Y''OZ''$, the image planes in which track images are formed by lenses L', L''.

For each lens the object distance from O is u and the image distance to O', O'' is v. The images of $P(x, y, z)$, P' and P'' are given by

$$P'(x', z'), \quad x' = \frac{vx}{u-y}, \quad z' = \frac{vz}{u-y},$$

$$P''(y'', z''), \quad y'' = \frac{vy}{u-x}, \quad z'' = \frac{vz}{u-x}.$$

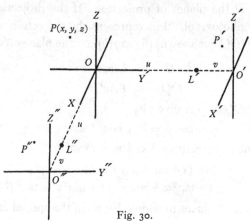

Fig. 30.

Thus
$$\left.\begin{aligned}
\frac{dx'}{dz'} &= \frac{(u-y)\,dx/dz + x\,dy/dz}{(u-y) + z\,dy/dz}, \\
\frac{dy''}{dz''} &= \frac{(u-x)\,dy/dz + y\,dx/dz}{(u-x) + z\,dx/dz}
\end{aligned}\right\} \tag{56}$$

and since
$$\left.\begin{aligned}
\tan\alpha &= \frac{dx}{dz}, \quad \tan a = \frac{dx'}{dz'}, \\
\tan\beta &= \frac{dy}{dz}, \quad \tan b = \frac{dy''}{dz''},
\end{aligned}\right\} \tag{57}$$

$$\left.\begin{aligned}
\tan a &= \frac{(u-y)\tan\alpha + x\tan\beta}{(u-y) + z\tan\beta}, \\
\tan b &= \frac{(u-x)\tan\beta + y\tan\alpha}{(u-x) + z\tan\alpha},
\end{aligned}\right\} \tag{58}$$

and so
$$\left.\begin{aligned}
\tan\alpha - \tan a &= \frac{m}{qp-mn}(q\tan b + n\tan a), \\
\tan\beta - \tan b &= \frac{n}{qp-mn}(p\tan a + m\tan b),
\end{aligned}\right\} \tag{59}$$

where
$$p = v - My, \quad q = v - Mx,$$
$$m = Mz \tan a - Mx, \tag{60}$$
$$n = Mz \tan b - My,$$

and $M = v/u$, the magnification of the photograph.

If we write $\Delta x' = (Mx - x')$, etc., and notice that in practice x, y, z are all small compared with u, then

$$\Delta x' = \Delta y'' = -\frac{x'y''}{v},$$
$$\Delta z' = -\frac{z'y''}{v}, \tag{61}$$
$$\Delta z'' = -\frac{z''x'}{v},$$

and q, p, m, n reduce to the forms

$$p = v - y''\left(1 - \frac{x'}{v}\right), \quad q = v - x'\left(1 - \frac{y''}{v}\right),$$
$$m = (z' \tan a - x')\left(1 - \frac{y''}{v}\right), \tag{62}$$
$$n = (z'' \tan b - y'')\left(1 - \frac{x'}{v}\right);$$

the corrections to rectangular projection are then derived from equations (59) and (62).

To effect the reduction from conical to rectangular projection, the points O', O'' must be located, and reference marks must be transferred to each photograph for the purpose. The corrections were shown by Blackett to be frequently many times the final precision of measurement. Using test objects which could be measured independently, he obtained a probable error of measurement by the methods outlined of about 10 min. of arc, although the corrections a to α might be as large as $5°$.

6.17. Deflexion of cosmic-ray tracks in plates

Measurements of the multiple scattering of cosmic-ray particles in plates are typical of angular measurements in the cosmic-ray field in which, for many purposes, the projected angle is sufficient and in which the technical problems concern the effect of chamber distortions, and the practical measurement of very small angles of deflexion in the presence of field curvature.

Chamber distortions

The causes of bulk movement of gas in the chamber have been treated in an earlier section (§ 6.6); in so far as they consist of a circulation, however, the effect on angular measurements is likely to be at least as serious as that in curvature determination.

If we write the distortion of a track in the form

$$\Delta(x) = \Delta_0 + \Delta_1 x + \Delta_2 x^2 + \Delta_3 x^3 + \dots, \qquad (63)$$

the coefficient Δ_0 is irrelevant in the methods of measurement we have considered, Δ_1 corresponds to an angular displacement, Δ_2 to the imposition of curvature and Δ_3 to the first non-circular distortion. If the gas was circulating about $x = 0$ with uniform angular velocity, Δ_1 alone would not be zero, and in all thermal circulation of gas in the chamber it is unlikely ever to be small compared with the other coefficients. The temperature control already outlined is thus of the first importance if angular measurements are to be made.

The equivalent operation to the inclusion in a set of momentum measurements of a series of tracks photographed in zero field is to photograph with a plate geometrically similar to the scattering layer but of negligible scattering power. This is not an altogether satisfactory procedure, for the thermal properties of the substitute plate cannot well be made to reproduce those of the real scatterer, thus vitiating the whole purpose of the exchange, while the operation of substitution of one plate for another is too tedious to permit sufficiently frequent return to the calibrating condition throughout a series of photographs. We are therefore led to place considerable stress on internal consistency as a guide to the systematic angular distortions.

To determine the systematic distortions we consider the algebraic sum of all measured deflexions for a selected group of tracks; this group is chosen to be the highest momentum group available, for these tracks the real scattering deflexions are as small as possible, and so the relative importance of chamber distortion is a maximum. If this sum is $\Sigma\theta$ over n tracks, $\Sigma\theta/n$ is taken as the systematic distortion (strictly, the systematic difference of distortion between the two halves of the chamber), and is applied as a correction to all measured deflexions. Further, if for the high momentum group we know that all measured deflexions are distortions, we can measure on this group the probable error of angular measurement. Blackett and Wilson (1938) quote the systematic distortion and probable error of distortion ($+0\cdot04° \pm 0\cdot16°$), based on tracks of measured momentum greater than 10^{10} eV./c. Since the expected mean scattering of a particle of momentum 10^{10} eV./c. in the plate concerned was $0\cdot09°$, and the maximum detectable momentum was only about 10^{10} eV./c., the uncertainty of measurement, $\pm 0\cdot16°$, is probably somewhat overestimated. If the relative displacements corresponding to this value are compared with those deduced from the uncertainty of momentum measurement in the same set of photographs, it is apparent that, as the mechanism of convection would suggest, the coefficient Δ_1 is distinctly greater than Δ_2.

6.18. The technique of deflexion measurements

The problem is to measure the angle of scattering between two parts of a track which do not meet and which are both curved. Clearly, the uncorrected angle between tangents at any two accessible points, A and B (fig. 31 (a)), is partly due to scattering and partly to the curvature imposed on the track in the magnetic field, and in practice the effect of the second factor is considerable. For example, in 1 cm. of lead, the mean angle of scattering is approximately $10^9/p$ degrees, while the change of direction over 1 cm. of path due to the field is $\dfrac{180}{\pi}\dfrac{300H}{p}$ degrees, that is,

$$\frac{1\cdot72 \times 10^4 H}{p} \text{ degrees.}$$

Hence, if $H = 10^4$ gauss, the correction for curvature is 17 % of the mean angle of scattering. In most practical cases it is likely to be larger; the scattering is less for materials of lower atomic number, for thicker plates scattering increases only as $t^{\frac12}$, while for accurate settings it is never satisfactory to determine the tangent at the actual point of contact at the plate; in typical measurements with lead, copper and gold plates by Blackett and Wilson, the correction varied from 30 to 40 %.

The measurements are conveniently made using a goniometer eyepiece fitted to a travelling microscope with traverses in two directions. The point on tracks at which the cross-wire of the goniometer is made tangential should be fixed so that location from track to track is straightforward. The sign of correction is clearly determined according as the scattering is with or against the curvature.

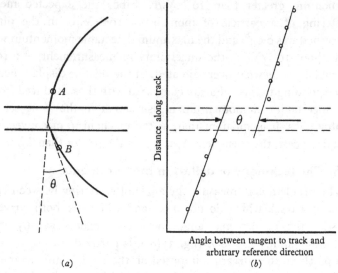

(a)   (b)

Fig. 31. The scattering of a fast cosmic-ray particle in a metal plate.

In a more exact method of measurement, the angle of the tangent to the track is plotted at intervals on either side of the plate (fig. 31 (b)), the plot on this scale of a uniform curve being a straight line of slope proportional to its curvature. Provided that the momentum of the particle is little changed in the plate (which is a condition for the correlation of observations of scattering with momentum) the two lines representing the two halves of the track will be approximately parallel, and the shift of one half relative to the other, representing the true scattering angle, is readily measured.

As for momentum measurements, the precision with which angles of scattering are observed depends on the quality of photograph. When this is high, individual settings of angle can readily be made to 0·1° and the mean value is thus considerably less than the uncertainty of distortion introduced by chamber disturbances.

6.19. Appendix: The production of magnetic fields

Interesting technical features are involved in the provision of magnetic fields necessary for the precision measurement of momentum in the cosmic-ray region; some of these features are discussed shortly here.

The field is produced either by use of a solenoid in air or by means of a magnet in which the field of the exciting coils is augmented by the presence of an iron yoke. The solenoid in air has a clear advantage in so far as the normal geometry of the cloud chamber, its mechanism behind and the camera axially in front is easily retained. When an iron core is introduced, however, although the chamber mechanism may reasonably be moved to the side of the apparatus, the necessity of photographing in a more or less axial direction remains, and a compromise has to be made between the ideal conditions of photography and an efficient magnetic circuit. Even when the usable volume of the cloud chamber is small, the geometry of photography makes a large air gap between the magnet poles inevitable.

The classical work of Kunze (1933a) and of Blackett and Occhialini (1933) was done with simple solenoids, and solenoids have also recently been described by French workers (for example, the aluminium coils described in the *Journal of Scientific Instruments*, **24**, 292 (1947)), for use with exceptional power facilities. In general, however, the performance of the simple solenoid can always be substantially improved by the addition of an iron circuit, *even if this is limited to the outer return path of flux only*. The important practical distinction now probably lies between the *electromagnet*, designed to give an air-gap field of the order of the saturation field in the magnet steel used, and the *iron-cored solenoid*, designed with much less iron arranged so as to interfere little with the freedom of arrangement of the cloud chamber, camera and associated gear, and giving a field appreciably smaller than the saturation field of the iron. It is, of course, in principle possible to build an electromagnet to provide approximately the saturation field over any required volume, but the ratio of weight and of cost between an electromagnet giving an operating field of 10,000–20,000 gauss and an iron-cored solenoid giving about 5000–7000 gauss over similar dimensions is great. In fig. 32, the electromagnet designed by Blackett in 1933, and now at Manchester University, is compared with an iron-cored solenoid now under construction for the laboratories at Manchester. The two magnets are drawn to the same scale, but while the electromagnet can cover a working volume of about 25 cm. diameter by 2 cm. deep at about 14,000 gauss, the iron-cored solenoid will cover a volume 55 cm. square by more than 10 cm. deep at from 5000 to 7000 gauss, at a comparable power input.

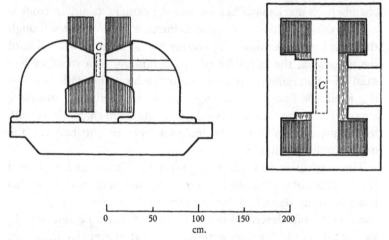

Fig. 32. Comparative dimensions of an electromagnet and an iron-cored solenoid of similar overall size and weight. Left, elevation of electromagnet; right, plan of iron-cored solenoid.

If the weight and cost are acceptable, the Blackett magnet represents very closely an optimum design provided that it is used in the neighbourhood of the correct gap spacing (we refer here in particular to the iron circuit; considerable variations of the exciting coils and of the cooling provision for them are possible which do not bear on the general argument). In practice, however, the governing design limitations are likely to be of cost or of weight; then the choice lies between a large field over a certain volume or a somewhat smaller field over a much larger volume. The relative values of these alternatives is a complicated function of the track geometry and absorber geometry of any particular experiment. If much space is occupied by absorbers, the intrinsic accuracy of momentum measurement may be greater in the low field, because of the greater length of particle trajectory available. It is not possible here to give definite criteria of suitability, since experimental requirements are now so varied.

The apparatus described by Brode (1939) is an iron-cored solenoid of the type which we have indicated; the Blackett magnet (1936) and that described by Jones and Hughes (1940) are electro-magnets of high efficiency. When exceptional power facilities are available, the distinction which we have used ceases to apply. Kunze (1933a), using power up to 500 kW., was able to reach a magnetic field of 18,000 gauss over a chamber of diameter 16 cm. without iron; similarly, the apparatus used by Anderson (1933) in much of the classical cosmic-ray work was essentially an iron-cored solenoid very heavily run (400 kW., 16,000 gauss, over 16 cm. diameter). These exceptional measures clearly lead to severe cooling problems, and the published data do not make it clear to what extent these have been solved.

CHAPTER VII

INTERPRETATION OF CLOUD-CHAMBER PHOTOGRAPHS

7.1. Introduction

In earlier chapters we have stressed the importance of the cloud chamber as an instrument of accurate measurement rather than of qualitative demonstration. Until recently most of these measurements were of a straightforward kind, although the technical difficulties of carrying them out might be considerable. Examples are the various particle collisions studied by Blackett (1922, 1923, 1925), the study of the $H^2 + H^2$ collision by Dee and Gilbert (1935), the search for ionization by neutrons by Dee (1932), and the various observations on ionization density as a function of momentum referred to in an earlier section (§ 2.5).

To an increasing degree, however, the cloud method has been strengthened by a technique of interpretation. Early examples are the arguments by which Anderson (1933) established the existence of the positron and later (N. & A. 1937) the distinction between particles in cosmic-ray showers and single unaccompanied cosmic-ray particles, while the application of the more highly developed technique of interpretation is excellently illustrated in the identification of a heavy meson ($\mu \sim 1000m_0$) by Rochester and Butler (1947).

In this chapter we shall attempt to cover some of the more important features of the technique of interpretation, and of the requirements which it imposes on the standards of photography and manipulation.

7.2. What can be observed in cloud-chamber photographs

(a) *Momentum*

The measurement of momentum has already been exhaustively examined (§§ 6.1–6.14); the requirements for momentum measurement are largely opposed to those for measurement of ionization density, and, since these are both quantities of the first importance in interpretation, the compromise between the ideal conditions for these measurements is discussed further below (§ 7.4).

(b) Density of ionization

This is ideally carried out by a drop count of average ionization up to a well-defined maximum cluster size (§ 2.2). On sharp tracks a method due to Williams (1939 c), based on the measurement of gaps between acts of primary ionization is also valuable, but this method is not suitable for use with counter-controlled tracks. With photographs of good quality in which, however, drop counting is not possible, a visual estimate of density may be attempted. When comparison tracks are present in comparable illumination this method can be of very considerable value, and an increase of 50 % in ionization density can possibly be detected. Without comparison material, it must be used with caution and can only be expected to yield very approximate estimates. Then a threefold ionization ratio would probably be noted.

(c) Production of secondaries in the gas

The secondary electrons along a cloud track can sometimes yield information of value. In the absence of magnetic field secondaries with sufficient energy to indicate the *direction of motion* of the primary particle are fairly frequent, but in a large field the initial direction of motion is observable only for rather energetic secondaries. The use of fast secondaries for determining the mass of the primary particle has been developed by Leprince Ringuet and his pupils; although material suitable for at all precise mass determinations is rare, secondary electrons fast enough to enable some limit to be placed on the *mass of the primary* are more common (see, for example, Rochester, Butler and Runcorn, 1947).

(d) Association in time

Apart from very exceptional cases, the resolving time of a controlling counter-array can be made so small that the chance that one or more of the counter-detected particles has come into association with the others at random is negligible. The resolving time of the chamber, however, is much longer (§ 5.3), and the chance that particles of abundant types (normal mesons, for example, or slow electrons), present in photographs and of an age indistinguishable from the counter-controlled group, may have come in at random is appreciable. Wilson (1939) estimated one random meson in

1000 photographs for a chamber of collecting area about 30 cm.^{2} and resolving time about 0·02 sec., and with the random particle entering in a solid angle restricted to about $\frac{1}{6}$ radian. In the chambers of much greater collecting area now used, which also probably have a rather greater resolving time, the chance of a random meson, indistinguishable from the counter-controlled particles, being present in a photograph is thus considerable. For many purposes, however, only the association of particles which come approximately from a single point is of importance. Then the effective collecting area for random particles is greatly reduced, and the chance of a random association again becomes small. In the extreme case of particles coming from a shower or explosion centre in a plate in the cloud chamber, it will always be quite negligible.

(e) Copunctuality: Reprojection

The question of strict copunctuality of tracks arises only in the gas of the cloud chamber; particles coming into the chamber must be expected to have suffered scattering in solid matter traversed, the probable value of which may or may not be known. If photographs in planes at right angles are available, the problem requires no comment; with a normal stereoscopic pair, however, reprojection is usually necessary to establish that tracks intersect to within a millimetre or less. Successful stereoscopic reprojection can best be achieved by using for the purpose the lens system and negative holder with which the photograph was initially taken, and it is important to design cloud-chamber cameras with this application in view.

(f) Recoil blobs

Scattering and similar processes when momentum is taken up by a nucleus are characterized by the recoil ionization of the heavy particle, when the recoil energy is large enough. These collisions are distinguished in this way from processes of spontaneous decay in which, of course, no recoil particle is involved.

7.3. Photographic quality demanded for interpretation

The necessity of measuring simultaneously in a single photograph both momentum and ionization density is the main reason for the stress which must now be placed on the maintenance of a technical level of photography in which every individual drop is photographed. Other features to which we have referred in the last section are also, of course, more effectively treated if this standard is reached. The track broadening on which the chamber resolving time is based is much less easily observed if drop clusters alone are recorded, and the approximate estimates of resolving time which have already been given are based on the assumption that individual drop images are recorded. The distinction between recoil clusters and the normal secondary electron clusters is also best appreciated if a continuous line of droplets come up to the recoil cluster from either side.

There is a further strong reason, however, for maintaining the standard of drop recording. In a clean chamber, the thin, even background of drops, if it is photographed, provides evidence that supersaturation and illumination from point to point in the chamber are sufficient for condensation on droplets and their subsequent photography. Whenever the trajectories of particles appear to begin (or end) in the gas, the question whether the trajectory in fact begins at that point or whether it goes on out of the light or into a region of low supersaturation arises, and the presence of background droplets in that region of the photograph is of the greatest importance (Rochester and Butler, 1947). Further, by reprojection methods it may be established to what distance beyond the apparent end of the trajectory the condition of adequate supersaturation and illumination extends.

All cloud-chamber photographs, but especially those in which the individual drops are recorded, are best viewed stereoscopically. The practice of taking a stereoscopic pair was first introduced so that reprojection might be possible if needed; while reprojection is still of importance, the value of the stereoscopic pair is certainly very much wider. In particular, the less dense levels of background cloud are extremely difficult to appreciate in a single photograph, but can be appreciated with confidence in stereoscopic viewing.

We shall discuss the rival claims of measurement of momentum and ionization in the following section. Even apart from this consideration, however, the standard of photographs suitable for exploratory interpretation is clearly formidable. The chamber must be clean and must give a thin but even background throughout its volume, being constantly maintained at an expansion ratio yielding substantially 100 % condensation on ions. Distortion or lack of supersaturation close to metal plates in the chamber must be avoided. In addition, expansion time and drop-growth time must be restricted as far as possible, since momentum must be measured to a high precision, perhaps on short or inclined tracks, while some concession is made towards ion counting.

It is, of course, true that a great deal of exploratory work is carried out with an indifferent chamber technique. It is also probable, however, that the extent to which the information obtained in this way falls short of what might reasonably have been attained is not always appreciated.

7.4. The momentum-ionization compromise

The particular relation between simultaneous measurements of momentum and ionization density, to which we have frequently referred, comes about because of the importance which now attaches to the identification of individual particles present in photographs and because the relation between momentum and ionization offers as a rule the most certain way of making this identification.

We discuss here the requirements which seem to be of importance at the present time. In principle, the relation between ionization density, particle momentum and particle mass is probably almost unique—at all momenta, any given ionization density and momentum correspond to two masses, of which general experience, over a very large part of the range of the momentum and ionization, excludes one. We have shown, however (§ 2.6), that the theoretical basis of the ionization loss at high momenta (roughly $pc > 10mc^2$) is not sufficiently well established to be used as a criterion of mass identification, while the existing measurements, even when made primarily as a study of ionization density (and not in the relatively stringent conditions which we are now discussing), are seriously discordant. Further, in the problem of the immediate future, the order of precision of momentum measurements which is required is fairly clear; momenta of the order of 10^9 eV./c. must be measurable to, say, $\pm 10 \%$, while the measurement of higher momenta can be correspondingly less accurate.

The immediate requirements, therefore, are for a maximum

detectable momentum of the order 10^{10} eV./c., under conditions when the low momentum rise of ionization can be measured. A measurement of ionization density to $\pm 10\%$ would represent a satisfactory standard.

In a shallow chamber working with argon at 1·5 atm., this momentum standard has been reached for 10 cm. of track in a field of 10^4 gauss (Wilson, 1938, 1939); more recently, Rochester and Butler* have reached a maximum detectable momentum in a deeper chamber of 8×10^9 eV./c. for similar conditions in a field of 7×10^3 gauss. This standard in a deeper chamber is almost certainly rather higher than that of the earlier work, and probably reflects the gain from a more developed thermal control of the chamber. It is possible that slight improvement could be obtained by reducing the growth time of drops. Neither piece of work quoted here, however, makes any concession to drop counting. In neither is it possible to resolve any but the smallest drop clusters, and drop counting is not possible, but the quality is sufficient for an increase of ionization density between 50 and 100% to be detected with certainty.

7.5. Lines of solution of the momentum-ionization compromise

Very little systematic work has yet been carried out to determine the conditions in which the measurement of both momentum and ionization can best be attempted to the order of precision indicated. We summarize here the main quantities upon the interrelations among which these conditions depend.

1. Precision of *momentum* measurement depends on the total time from traversal to photography (normally dominant), and on the uncertainty of the line defined by a finite number of diffused drops.

2. The total time consists of expansion time (traversal to full supersaturation) + growth time (to photography).

    2.1. Expansion time may be reduced by further mechanical developments, the use of a light gas (H_2) for driving the piston, the use of a low expansion ratio.

    2.2. Growth time is reduced by the use of light gases, the use of a vapour of high rate of drop growth (water better than alcohol, better than 75/25 alcohol-water).

* Unpublished.

3. The uncertainty of momentum measurement because of diffusion is normally of secondary importance; it may be increased, and, other things equal, might become dominant if a gas were used in which the vapour has a high coefficient of diffusion, and if total ionization is small (i.e. the light gases).

4. Ionization counts to a precision of the order 10 % depend mainly on the resolution of small clusters, that is by the use of an emulsion of high resolution, high magnification, large diffusion, small total ionization, and by a high standard of chamber cleanness (a large tolerable expansion range in which substantially 100 % condensation takes place).

(*The conditions in which drop counting may be carried out on counter-controlled tracks have been examined by C. O. Green (private communication) from the point of view of the identification of an individual particle when a given (short) length of track only is available. While the statistical accuracy inherent in the number of primary ionizations can clearly not be exceeded, it is not possible in a counter-controlled track to identify primary ionizations, and the smallest identifiable cluster increases in size with increasing diffusion of the track. Moreover, the fraction of track which is rejected, if clusters of greater than given size are to be excluded, also increases with increasing diffusion. Green concludes that the attainable statistical accuracy of the drop count is only a slowly varying function of the maximum size of cluster included in the count, and that accordingly the stress which has been placed here on the need for additional diffusion in order that clusters of moderate size may be resolved is not well founded. It seems probable that, provided individual drop images are recorded, the normal track diffusion of counter control will allow ionization estimates which differ little in precision from the optimum, although the accuracy will still be significantly below that defined by primary ionization.)

7.6. Characteristic features of particle behaviour used for purposes of interpretation

The central problem of the interpretation of exploratory photographs is the recognition of the particles involved in a particular

* Added November 1949.

event. Ideally, particles are recognized by simultaneous observation of momentum and ionization density, but this is only possible over a restricted momentum range, and more often than not the particles in which we are interested do not fall into this range. We now consider other features which can give indications as to the identity of particles.

(a) The relativistic transformation of collisions and decay processes

Whether or not the geometry of an event is known in the centre of mass frame of reference with respect to the direction of the transformation, an estimate of the magnitude of the transformation can very frequently be made. This is mainly of value when the momentum of some at least of the particles concerned is known, preferably by direct measurement, but if this is not possible, by indirect methods ((b), (c) below). The formation of electron pairs is particularly characteristic; for example, the pair-like event in the gas in fig. 1 of Rochester and Butler (1947) cannot be an electron pair, since the momenta of both particles is certainly greater than 10^8 eV./c. and the angle between them is $60°$.

(b) The free-electron collision

The observation of a fast secondary electron produced by a primary particle of known momentum leads in principle to a mass determination of the primary. Its application as setting a limit of mass is quite often possible. It has been shown (Gorodetsky, 1942) that for the important conditions

$$\left(\frac{\mu}{m}\right)^2 = \left(\frac{pc}{mc^2}\right)^2 \left\{\left(\frac{p' \cos \theta}{p_c}\right)^2 - 1\right\}, \qquad (64)$$

where μ, p are the mass and momentum of the primary, m, p' of the electron, θ is the angle of projection of the secondary, and

$$p_c c = \{(mc^2)^2 + (p'c)^2\}^{\frac{1}{2}} - mc^2.$$

The value of this expression lies mainly under conditions in which θ is small, and the difference of $\cos \theta$ from unity is rarely known significantly. In a photograph discussed by Rochester, Butler and Runcorn (1947), a particle having $p \sim 2 \times 10^7$ eV./c. leaving a lead plate is accompanied by a slow electron of $p' \sim 1\cdot5 \times 10^5$ eV./c. Since observation of θ must be regarded as completely masked by scattering, and p' as measured is a lower limit to the momentum at formation, an upper limit of the primary mass alone is possible, $\mu < 300m$. This value is consistent with an ionization-momentum estimate, and affords confirmatory evidence that the particle is the normal cosmic-ray meson of mass ~ 200 m. The detection by this method of a particle of mass $\sim 1000m$ has been described in detail by Leprince Ringuet and his co-workers (1946), and the limitations of the method have been studied critically by Bethe (1946).

(c) Scattering in plates

The scattering of a fast particle passing through a metal plate may be considered as made up of the multiple Coulomb scattering arising from extranuclear collisions in the material traversed together with the single, or at most plural, large-angle scattering in close collisions with individual nucleons within the nuclei of the material. For cosmic-ray mesons it has been established (Blackett and Wilson, 1938; Wilson, 1940; Code, 1941) that the occurrence of large single scattering in plates of the thickness which is accommodated in cloud chambers is extremely rare, and can for our purpose be ignored. Then, with only multiple Coulomb scattering effective, some indication of particle momentum is possible $(\bar{\theta} \propto p/\beta)$ from the scattering of a particle in a plate, and since the magnitude and distribution of Coulomb scattering is accurately known, the indications can be expressed precisely. However, it has not yet been established for other categories of particle that the single scattering in such plates is small; indeed, there is indication that for some particles at least it is not. It therefore appears that *small* scattering may be taken for all particles as an indication of high momentum, while *large* scattering of particles of *high* momentum is probably characteristic of certain particles, although precisely what particles is not yet known.

(d) Characteristics of electron cascades in plates

It is well known that some at least of the events in which many particles appear to be formed in a metal plate in the cloud chamber are not electron cascades. The importance of these rather rare happenings makes the recognition of the characteristic features of the normal cascades a matter of concern.

Except in extreme cases, little is to be learned from the mere number of particles in the event, the characteristic features of a cascade being the localized apparent region of origin (which may be obscured, however, if several cascades have been initiated at small separations), the distribution of momentum among the particles and their angular distribution. Since the main angular dispersion of a cascade is due to scattering, the angular distribution and the momentum distribution are related.

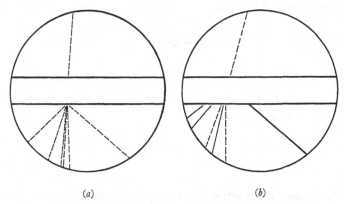

(a)   (b)

Fig. 33. Characteristic appearance of an electron cascade (a) and of a nuclear disruption with the emission of heavy particles (b), from a plate of high Z in the cloud chamber.

A cascade leaves a metal plate within an area of diameter of the order of the cascade unit of length; since this unit is about 0·4 cm. for lead, the area of emergence from lead or from other heavy metals is almost point-like. In this respect, a cascade differs clearly from the emission of heavy particles for which scattering is of secondary importance and the angular distribution is determined by the relativistic transformation of the collision process. For the emission of heavy particles (see, for example, Rochester, Butler and Runcorn (1947), fig. 1), the track may leave the plate over an extended area, and, subject to the uncertainties of scattering, project back to a point in the metal plate (fig. 33). The presence of non-electronic particles in an otherwise normal cascade is also shown by this property.

The momentum distribution of electrons in a cascade has been treated exhaustively, and standard works such as Janossy (1948) may be consulted. The features are of two kinds: those common to all showers, and those which are a feature of the particular plate thickness in use, and also of the energy of the incident electron or photon. Except for very undeveloped showers, the most generally characteristic feature of the distribution is the uniformity of momentum, a large fraction of the outcoming electrons will be of momentum between 10^7 and 10^8 eV./c., for it is important to notice that the particles of still lower momentum are very largely removed by scattering. Table XVII gives figures for the outcoming particles from a gold plate of about 8 cascade units thickness, which is typical of the thickness of plate now widely used (Wilson, 1939).

TABLE XVII. *Cascade development in 2 cm. gold*

Momentum of incident electron	4×10^8 eV./c.	10^9 eV./c.	2×10^9 eV./c.
Probable number of secondaries $> 10^7$ eV./c.	2	6	12
Probable number of secondaries $> 10^8$ eV./c.	$\leqslant 0.001$	< 0.1	0.5

The angular distribution arising from scattering is characterized by the confinement of energetic electrons (which can have travelled only a distance of the order of the cascade unit, and must previously have come of yet more energetic parentage) to a narrow central core, and energetic particles ($p > 10^8$ eV./c. for lead) which emerge at large angles ($\theta > 20°$) are probably not electrons.

(e) *Probability arguments applied to the distinction between decay and collision processes*

In the interpretation of recent photographs, Rochester and Butler (1947) use a probability argument for considering particular events happening in the gas of the chamber as spontaneous decays rather than as collision processes. Although in these examples there is contributory evidence which at least narrows the kinds of collision which are conceivable, the probability argument is a crucial one. It is based on the fact that spontaneous decay is unaffected by the presence of matter, and hence is ten times more probable in 30 cm. of gas in the chamber than in 3 cm. of metal plate, while for any reasonable interaction, at an energy which is much too high for any specific effect to be in question, a collision process will be some hundreds of times more probable in the metal plate than in the gas. If, then, the event would be recognizable by the particles emerging from a plate, and if the particles are of a kind and of an energy able to get out of the plate (which, in the example quoted, one certainly is, the others probably are), the occurrence of even an isolated example in the gas without many recognizable examples from the plate is not consistent with the occurrence of a collision. (Normal pair production, for example, shows the expected kind of relation between production in a plate and in the gas of the chamber.)

REFERENCES

ADAMS, ANDERSON, LLOYD, RAU and SAXENA (1948). *Rev. Mod. Phys.* **20**, 334.
AITKEN (1880–1). *Collected Papers*, 1923, p. 34. Cambridge.
AITKEN (1911). *Collected Papers*, 1923, p. 495. Cambridge.
ANDERSON (1933). *Phys. Rev.* **43**, 491.
AUGER and DAUDIN (1945). *J. Phys. Radium*, **6**, 302.
BARRETT and GERMAIN (1947). *Rev. Sci. Instrum.* **18**, 84.
BEARDEN (1935). *Rev. Sci. Instrum.* **6**, 256.
BECK (1941). *Rev. Sci. Instrum.* **12**, 602.
BECKER and DORING (1935). *Ann. Phys., Lpz.*, **24**, 719.
BETHE (1930). *Ann. Phys., Lpz.*, **5**, 325.
BETHE (1946). *Phys. Rev.* **70**, 821.
BLACKETT (1922). *Proc. Roy. Soc.* A, **102**, 294.
BLACKETT (1923). *Proc. Roy. Soc.* A, **103**, 62.
BLACKETT (1925). *Proc. Roy. Soc.* A, **107**, 349.
BLACKETT (1934). *Proc. Roy. Soc.* A, **146**, 281.
BLACKETT (1936). *Proc. Roy. Soc.* A, **154**, 564.
BLACKETT (1937). *Proc. Roy. Soc.* A, **159**, 1.
BLACKETT and OCCHIALINI (1933). *Proc. Roy. Soc.* A, **139**, 699.
BLACKETT and WILSON (1937). *Proc. Roy. Soc.* A, **160**, 304.
BLACKETT and WILSON (1938). *Proc. Roy. Soc.* A, **165**, 290.
BLOCH (1933). *Ann. Phys., Lpz.*, **16**, 285.
BOHR (1913). *Phil. Mag.* **25**, 10.
BOHR (1915). *Phil. Mag.* **30**, 581.
BRODE (1939). *Rev. Mod. Phys.* **11**, 222.
CERENKOV (1937). *C.R. Acad. Sci. U.R.S.S.* **14**, 101.
CHAUDHURI (1948). *Nature, Lond.*, **161**, 680.
COCCONI, LOVERDO and TONGIORGI (1946). *Phys. Rev.* **70**, 846.
CODE (1941). *Phys. Rev.* **59**, 229.
CORSON and BRODE (1938). *Phys. Rev.* **53**, 773.
COULIER (1875). *J. Pharm. Chim., Paris*, **22**, 165.
CWILONG (1947). *Proc. Roy. Soc.* A, **190**, 137.
DAUDIN (1943). *Ann. Phys., Paris*, **18**, 145, 217.
DAUDIN (1947). *J. Phys. Radium*, **8**, 301.
DEE (1932). *Proc. Roy. Soc.* A, **136**, 727.
DEE and GILBERT (1935). *Proc. Roy. Soc.* A, **149**, 200.
ENDT (1948). *Physica*, **14**, 97.
FARKAS (1927). *Z. phys. Chem.* **125**, 236.
FERMI (1940). *Phys. Rev.* **57**, 485.
FLOOD (1934). *Z. phys. Chem.* **170**, 286.
FRENKEL (1946). *Kinetic Theory of Liquids*. Oxford.
FRISCH (1935). *Naturwissenschaften*, **23**, 166.
GLOSIOS (1939). *Ann. Phys., Lpz.*, **34**, 446.
GORODETSKY (1942). Thesis, Paris.
HALPERN and HALL (1940). *Phys. Rev.* **57**, 459.

HALPERN and HALL (1948). *Phys. Rev.* **73**, 477.
HAZEN (1942). *Rev. Sci. Instrum.* **13**, 247.
HAZEN (1943). *Phys. Rev.* **63**, 107.
HAZEN (1944). *Phys. Rev.* **65**, 259.
V. HEERDEN (1945). Diss. Utrecht.
HERZOG (1935). *J. Sci. Instrum.* **12**, 153.
JANOSSY (1948). *Cosmic Rays.* Oxford.
JANOSSY and BROADBENT (1948). *Proc. Roy. Soc.* A, **192**, 364.
JOLIOT (1934). *J. Phys. Radium*, **5**, 216.
JONES and HUGHES (1940). *Rev. Sci. Instrum.* **11**, 79.
KUNZE (1933*a*). *Z. Phys.* **83**, 1.
KUNZE (1933*b*). *Z. Phys.* **80**, 559.
LANGSDORF (1939). *Rev. Sci. Instrum.* **10**, 91.
LEPRINCE-RINGUET and LHÉRITIER (1946). *J. Phys. Radium*, **7**, 65.
LOCHER (1937). *Franklin Inst. J.* **224**, 555.
MEYER LEIBNITZ (1939). *Z. Phys.* **112**, 569.
MIE (1908). *Ann. Phys., Lpz.*, **25**, 377.
MILATZ and VAN HEERDEN (1947). *Physica*, **13**, 21.
NEDDERMEYER and ANDERSON (1937). *Phys. Rev.* **51**, 884.
POWELL (1928). *Proc. Roy. Soc.* A, **119**, 553.
RAYLEIGH (1910). *Proc. Roy. Soc.* A, **84**, 25.
ROCHESTER and BUTLER (1947). *Nature, Lond.*, **160**, 855.
ROCHESTER, BUTLER and RUNCORN (1947). *Nature, Lond.*, **159**, 227.
SCHARRER (1939). *Ann. Phys., Lpz.*, **35**, 619.
SEN GUPTA (1943). *Proc. Nat. Inst. Sci. Ind.* **9**, 295.
SHIMIZU (1921). *Proc. Roy. Soc.* A, **99**, 425.
SKROMSTAD and LOUGHBRIDGE (1936). *Phys. Rev.* **50**, 677.
THOMSON (1888). *Applications of Dynamics to Physics and Chemistry.* London: Macmillan.
TOHMFER and VOLMER (1938). *Ann. Phys., Lpz.*, **33**, 109.
VOLMER and FLOOD (1934). *Z. phys. Chem.* **170**, 273.
VOLMER and WEBER (1926). *Z. phys. Chem.* **119**, 277.
WEBB (1935). *Phil. Mag.* **19**, 927.
WILLIAMS (1931). *Proc. Roy. Soc.* A, **130**, 328.
WILLIAMS (1939*a*). *Proc. Camb. Phil. Soc.* **35**, 512.
WILLIAMS (1939*b*). *Proc. Roy. Soc.* A, **169**, 531.
WILLIAMS (1939*c*). *Proc. Roy. Soc.* A, **172**, 194.
WILLIAMS (1940). *Phys. Rev.* **58**, 292.
WILLIAMS and TERROUX (1930). *Proc. Roy. Soc.* A, **126**, 289.
WILSON, C. T. R. (1897). *Philos. Trans.* **189**, 265.
WILSON, C. T. R. (1899). *Philos. Trans.* **192**, 403.
WILSON, C. T. R. (1912). *Proc. Roy. Soc.* A, **87**, 277.
WILSON, C. T. R. (1923). *Proc. Roy. Soc.* A, **104**, 1, 192.
WILSON, C. T. R. (1933). *Proc. Roy. Soc.* A, **142**, 88.
WILSON, C. T. R. and WILSON, J. G. (1935). *Proc. Roy. Soc.* A, **148**, 523.
WILSON, J. G. (1938). *Proc. Roy. Soc.* A, **166**, 482.
WILSON, J. G. (1939). *Proc. Roy. Soc.* A, **172**, 517.
WILSON, J. G. (1940). *Proc. Roy. Soc.* A, **174**, 73.
WILSON, R. R. (1941). *Phys. Rev.* **60**, 749.

INDEX

Printed in the United States
By Bookmasters